Combat Trauma-
PTS(D)
The Final Edition

By Frederick W. Nolen, Ph.D.
Licensed Psychologist

Please contact me via frednolen@msn.com if you have questions or requests for additional consultations, workshop presentations or program development for PTSD.

ISBN
978-1-958400-29-6

Last revision, 10 February, 2023

TABLE of CONTENTS

...

You can also freely share any and all of this information (as long as you give credit) to any combat vet, spouse, child or family member of a combat vet for as long as any of us shall live. **Their spouses and family are probably the ones that will help them the most. They all need to know the information in this article and how to apply it.**

I've had many combat veterans tell me "I wish I'd known about this 30 years ago." I wish they'd known about it 30 (now 40) years ago, too. I wish they'd known about it too, given what they had all gone through in the unconscious grip of their past. Help our new men and women combat vets out…spread the word!

Caveat 1: This treatment information is NOT meant to be automatically applicable for the severely head-injured soldier. The location and severity of their head injury must be considered to evaluate the soldier's diminished capacity for cognitive processing and impulse control from the head injury.

> There are some serious brain damage issues that need to be scientifically evaluated by their caretakers. The hallmark of brain damage is lack of impulse control. The "signature" injury of the Iraqi/Afghanistan Wars, Traumatic Brain Injury (TBI) is the brain injury from IEDs and RPGs. Add those injuries to the proximity and emotional bonding of the soldier to their weapon and you have the setup for the greatest of all back-home tragedies: suicide and homicide. (See more in the Traumatic Brain Injury section, below).

Caveat 2: This information is most applicable for outpatient treatment and psychoeducation.

Caveat 3: When I write "he", I also mean she, too.

Caveat 4: I warn the reader that other combat-exposed personnel (nurses, doctors, medics) often have it, too. The healers often need healing. Many of them tell me one case always "got them" even thought they were professionally numb to the maiming, moaning, blood and bleeding. That case often, but not always, involved children or women.

Caveat 5: I give many horrific but true examples of what those in the fog of war do and experience. This is not for the weak of stomach. However, if you have a visceral reaction to reading about the events, just try to empathize how much more emotional it was for those who directly experienced it.

Caveat 6: This book contains many "worst case" scenarios" for physical,
mental and family problems from wounded combat veterans. Not all veterans come back and have such severe problems.

Caveat 7: Many in the military petitioned for the "D" to be removed from PTSD of the DSM5. This was based on the attitude that it was not a disorder or disease but a natural response to the human mind and body for the horrendous carnage experienced in combat. I agree with this attitude.
The DSM5 committee did not agree. Therefore the "D" stayed in PTSD. I heartily agree with those who wished the D be removed. I think it should be called PTSR (Post Traumatic Stress Response). PTSR is NOT a disorder. It is an unavoidable, uncontrollable, physical and mental reaction to carnage, especially when it is intentionally done by one person to another. The uncontrollable physical reaction part of it is mostly predictable given the physical aspects of the traumatizing situation. For example, if somebody calls you an insulting name, you will have a physical reaction you can't control, but it will be mild if the previous insults to you around this word were not physical. If they throw an ink pen and hit you with it, you will have a greater physical reaction and you cannot control it. If they hit you with a laptop computer, you will have a bigger physical reaction that you cannot stop. If a 500-pound bomb goes off a block away, you will have an even bigger physical reaction that you cannot stop. The tricky thing about PTSD is the physical reactions continue to occur, uncontrollably (at first), as if the attack is happening again. These physical reactions from then-and-there get "triggered" by here-and-now events. See the Triggers discussion later in this book.

...

Another factor that determines one's reaction to death is "Is this a 'good' death". All societies have a fantasy about a "good death". We all will die. We all just want it to be a good death by meeting the following criteria:

a) We die when we are old.
b) We die suddenly.
c) We die painlessly.
d) We die with a completed "bucket list".
e) We die without disfiguration.
f) We die peacefully.
g) We die with our loved ones around our bed, boosting us on to the next level, whatever that maybe.

Death in combat usually violates all those aspects of the "good" death. The combat soldier is usually, young, fit, dies mangled if not obliterated, painfully (often crying out for their mother God, or both), screaming in agony, alone.

Caveat 8: this is the last edition I will do of this book. Therefore, I include the horrors of war that are inflicted on the innocent children our soldiers tried to help and the women who were "in bed with" (had sex with) the enemy, the American soldiers. The instances I write about happened in Vietnam and OIF/OEF. I am sure they have been many more atrocities; I have sure they have happened in many other wars. These are in very end of the Trauma Triggers section of this book in a section titled Atrocities to Women and Children.

HISTORY OF THE PHENOMENON

Planet earth has suffered the short-term and long-term effects of war for the entire history of mankind. There's documents and documentaries. There's even a war channel on American television. It used to be called "The History Channel" for some ironic reason. Recently they appropriately renamed it the "Military Channel".

Unfortunately, the world has known or been taught much less about the effects of war on the soldier (besides the obvious...they make it back in one piece or they don't). American folklore has had different names for the effects of combat trauma for centuries. People called it "The Reverie" after the Civil War, "Shell Shock" and "The Thousand Yard Stare" for World War I veterans, "Combat Fatigue" for World War II veterans, "Vietnam Vet Syndrome: and Post Traumatic Stress Disorder (PTSD) since the DSM-III came out in 1980.

The American military tried to re-label it "combat stress" during the early part of Operation Enduring Freedom/Operation Iraqi Freedom (OEF/OIF) but it didn't stick. Although the literature and media often portray these different labels for it as synonymous, they are not exactly identical.

I realized this while watching a documentary on WWI. It showed film clips of a soldier suffering from what they then called "shell shock". It showed a man so jittery he couldn't stop from having grand mal- type clonic spasms of his voluntary muscles on top of severe shaking just sitting down.

This high level of involuntary twitching and spasms of the usually-voluntary muscles was caused by the constant and extreme emotional and physical tension of being shelled by enemy artillery. It was constant, daily, thousands-of-shells-at-a-time for weeks at a time. The lethality and dangerousness of it was emphasized by the average life-span of a soldier in the trenches: six weeks!

I have never heard or seen the same type of spasms in WWII veterans or later because they weren't exposed to the constant, daily shelling of the trench warfare until I met an American, 94-year-old WWII combat medic in 2013. I will tell you more about him later.

Then there's the "Thousand Yard Stare". What is the "Thousand Yard Stare"?

I have seen it trivialized in popular civilian literature for many decades. They tend to equate it with staring off into space or spacing out. For example, Dale Brown wrote: Patrick was silent for a few moments, adopting his infamous "**thousand-yard stare**" as his mind turned over possibilities. ..." (Strike Force: A Novel, Publisher unknown, p 183).

It's not that at all. Those with the "Thousand Yard Stare" don't "adopt it" any more than fish "adopt" gills.

Charles Henderson also wrote: He could see beginnings of the telltale one-**thousand-yard stare**, the stoic expression on a face that had seen its share…" (p 35).

That's wrong, too. It's not a stoic expression at all, either.

I have seen it in severely abused children and adults. I've also seen it in pictures of Holocaust and Bataan Death March survivors.

"This is the thousand yard stare"

They can be looking at you but their eyes are hollow. They see you but you don't fill their eyes at all. Their eyes are empty. They can be following your movements with their eyes but they're not all here. They will talk in response to your words, but they aren't just talking to you. They are talking to the words. They aren't spacing out. They simply aren't all here. The personal part(s) of their mind are hiding. The missing part(s) only come back out of hiding when they think they are safe.

We can expound on this phenomenon (One thousand Yard stare) further as a symptom of PTSD. The explanation you have given above is about what exactly it is for the person exhibiting this symptom. We can add here an example (experience) of the people whose closed ones suffered from it (just to make them understand this whole thing a little better). Here is an example:

"The thousand-yard stare or two-thousand-yard stare is the unfocused gaze of a battle-weary soldier. The stare is a characteristic combat stress reaction which may be a precursor to, or symptom of, post-traumatic stress disorder.

...

The phrase was popularized when, in 1944, Life magazine published the painting Marines Call It That 2,000 Yard Stare by its World War II artist and correspondent Tom Lea. The painting was a portrait of a young marine at the Battle of Peleliu in 1944, and is now held by U.S. Army Center of Military History, Fort Lesley J. McNair, Washington, D.C. About the real-life Marine who was his subject, Lea wrote:
"He left the States 31 months ago. He was wounded in his first campaign. He has had tropical diseases. He half-sleeps at night and gouges Japs out of holes all day. Two-thirds of his company has been killed or wounded. He will return to attack this morning. How much can a human being endure?"

Frank Johnston, a Vietnam War photographer, used the term in a 2001 *Smithsonian* magazine interview. Johnston says, "I looked up and saw a Marine with what they call the thousand-yard stare, and I lifted my Leica and snapped his picture. The soldier's gaze never left my lens."

When recounting his arrival in Vietnam in **1965**, then-Corporal Joe Houle said he saw no emotion in the eyes of his new squad: "The look in their eyes was like the life was sucked out of them." Later learning that the term for their condition was *the 1,000-yard stare*, Houle acquiesced, "After I lost my first friend, I felt it was best to be detached."

https://psychology.fandom.com/wiki/Thousand-yard_stare

The popular depiction of "shell shock" in post-WWII movies showed the spaced-out (but sniveling) soldier lying in a hospital bed or in a wheel chair with a bandage around his head. The cure was to give him a pep talk, guilt trip him, try (unsuccessfully) to convince him he was just feeling sorry for himself, slap him around and ship him back to the front. He wasn't really hurt…he just needed a kick in the pants.

I actually saw this in a post WWII movie but I can't remember the name of it or the actors. It displayed the old notion, "If you fall off a horse, the cure is to put you right back up on the horse and show him who is boss". It was, "Patch 'em up; ship 'em back". The WWI military called it the "PIE" method (proximity to the battle, immediacy of treatment and expectancy of recovery, including return to duty).

The DSM-III PTSD criteria were heavily loaded toward combat trauma sequellae. The application of these criteria to natural disasters and rape victims followed later after that in the DSM-IV. I believe PTSD became an accepted combat-induced trauma because the Vietnam Veterans of America association (now called Veterans for America), among other veterans' groups, were heavily influential in getting the diagnosis officially recognized by the AMA (American Medical Association) and WHO (World Health Organization).The DSM-IV criteria were modified to be applicable to other trauma survivors (from natural disasters, rape and sexual abuse), but for some reason eliminated Survivor Guilt and other important components relevant to most combat veterans.

The DSM5 version is little changed other than the entire diagnosis has been put in a "Trauma-and Stressors-related Disorders" category by itself instead of being in the Anxiety Disorder section.

THE AMERICAN MYTH ABOUT WAR

Before I examine the science of combat trauma, I want to expose and analyze the American myth about the reality of war. It is exemplified by post WWII movies such as "The Longest Day", all the other post- WWII movies, Vietnam era movies such as "Deer Slayer" and "Platoon". The rarities are such films as the Vietnam-era "Casualties of War" with Sean Penn and Michael J. Fox (rape and murder of Vietnamese civilian female) or "Cease Fire (with Don Johnson, 1985), Saving Private Ryan, We Were Brothers, that show glimpses of the ignobility (rape and murder of civilians), futility and helplessness that is so frequently the reality of war, in war and back home, then and now.

I grew up watching "The Longest Day" or "The Dirty Dozen, or the hundreds of other typical Hollywood post-WWII "the-good-guys-kill-the-bad-guys-without-getting-as-much-as-a-scratch" war pictures. "Platoon" kept the fantasy going for Vietnam War junkies as much as possible (except one good guy kills another good guy).

After watching the typical post-WWII movies, I remember I would play-act "storming the machine gun nest" with my brother. I was seven; he was eight. I even thought that if you ran zigzagged you could dodge the bullets. I did, really!

Here's the myth in slow motion. The good guys are good looking, have all the cool gear (even called "sexy" by some recent, real American military staff), kill the strange-looking enemy with magnificent and noble shots (one shot, one-kill) and never even get a scratch. You only shoot the enemy. The direct hit is what gets you. If you die, you have this sad, farewell discussion with your best buddy. You die with him holding you. You are such a good shot you can shoot the gun out of the enemy's hand and subdue him nobly.

The screams and explosions are all under 90 decibels. The screams are made consciously, forcing the air out of their lungs as hard as the actor can. The action ends when the enemy surrenders. The soldier stays young and virile forever.

They go back home, get the girl, get the good job, make babies and live "happily ever after".

Here's some of the reality in slow motion (No, I'm not claiming I am a combat veteran. Ask a combat vet if you're really curious. Good luck if he tells you anything.):

The good guys are good looking until they take their uniforms off…then they look like average dudes. Our guys do have the coolest, most sophisticated combat gear on the planet…but their guys kill our guys with feces-covered sticks, WWI rifles, WWII bombs buried by the roadside, guns and ammunition we supplied their leader 30 years ago because he said he'd be a democratic ruler, or a box cutter. I knew a Vietnam veteran who saw an old Viet Cong man shoot down an American helicopter with one round from a single-shot, bolt action WWI rifle. Fifteen-million-dollar helicopter vs $15 rifle: The rifle won.

Many of the good guys get wounded and suffer forever, both physically and emotionally.

Sometimes you accidentally shoot and/or bomb your own guys (friendly fire).

The concussion of a bomb going off 100 yards away can blow your intestines out of your body or make you deaf forever. I've never seen a measure (in decibels) of a bomb or artillery or IUD blast. You don't merely hear them. The sound goes through your entire body. You feel them, too.

You gotta "hit the dirt" just right during an air-raid or artillery barrage or the concussion of the blast will transmit through the earth and jellify your intestines.

The more current, increasingly detailed, supposedly more-lifelike, slow motion movie shots-hitting-the-soldier (eg, "Platoon" or "Band of Brothers") always show the blood spurting from the shoulder or head or wherever. They rarely show the arm being blown completely off, the head being blown completely off, the eyes being blown out of the socket, decapitated heads flying off and killing other soldiers, flesh melting from napalm or heat of explosions. Special effects people either don't know about real wounds or can't imitate them exactly. Trust me, they would if they could. They are certainly trying.

The screams of the severely or mortally wounded are impossible to intentionally imitate. The air is involuntarily wrenched out of their lungs causing sounds men and women cannot imitate…ever. Men scream like rabbits scream when they are getting mauled. Most soldiers die crying for their mothers. They cry out for or curse God.

The only smells you get watching the war movie are the popcorn, soda, Gummie Bears, candy bars and perfume. You don't smell the blend of sweat, urine, hot blood and feces that men eject when they die or get so scared they lose body control.

You try to shoot the enemy's gun out of his hand (to mercifully and nobly disarm him) but you shoot his hand instead. The super-cool, maximum-lethal round you use in your super-cool weapon tumbles just like it was designed to tumble. Its tumble maximizes its kill potential. The effect of the tumble whips his (or her) hand and arm around, hitting him (or her) in the head, killing him (or her) by crushing their face or skull.

Then you wonder, "Where is God today?" and you puke.

Yes, you'll find pictures of their families in their pockets. You are horrified and maybe feel ashamed.

You are horrified. You vomit. You wonder where God is today. You were just trying to nobly wound him/her. You're the good guy. God is on your side. Right?

When you go back home, you may not get the good job… or get the good job you gave up when you were called up.

You may not get the brass band and parade.

You may or may not get the girl. If you had the girl, she may have been a "GI Jodie" (the WWII term for a girl who cheats on her husband-boyfriend-soldier when he is away at war). She may or may not be around when you come back.

You may be such an emotional, drug/alcohol abusing wreck that she and the kids don't stay around forever if they are there when you return.

Part of the fantasy is what most American soldiers have when they sign up for the military. The training they get prepares them a little bit for the realities of war. (For example, they now use silhouette targets in basic training (boot camp) for target practice. During WWII, they used "bullseyes" and it was estimated by the Department of Defense that only 5% of the armed soldiers in any group were actually shooting at the enemy to kill them. That "effective firepower" percentage went up to 60% during Vietnam, I was told, thanks to the silhouettes of human profiles used for target practice in boot camp.

That "effective firepower" ratio would go up even more if they used videos to train the troops, now. Oops, I forgot, they are doing that now. They just call them video "games" (not "training you to kill" games). They are available in your nearest video store or gamer outlets or on the internet: Call of Duty, Ghost Recon, Battlefield and SOCOM, to name a few.

Nothing can dispel the fantasies completely except war itself. Factor in everyone's illusion of invincibility and bullet-proof-ness, their fantasy of being protected by God, their illusion that bad things don't happen to good people and by then, it's too late.

They are soldiers in the "fog" of war and, maybe…survivors.

...

DSM- 5 POST-TRAUMATIC STRESS DISORDER (PTSD)

DIAGNOSTIC CRITERIA

The following are the DSM-5 diagnostic criteria for PTSD. They differ very little from the DSM-IV-TR criteria. I want to detail the many components of PTSD discussed in the DSM-5 because they are truly applicable and predictable **sequellae** (aftermath) of many combat veterans. I urge you to remember the strong positive correlation (relationship) between the amount of tissue trauma experienced (inflicted on others, experienced to oneself or witnessed to your comrades) and psychological trauma.

A. Exposure to actual or threatened death, serious injury, or sexual violence in one or more of the following ways:

(1) Directly experiencing the traumatic event(s).
(2) Witnessing, in person, the event(s) as it occurred to others.
(3) Learning the traumatic event(s) occurred to a close family member or close friend.
(4) Experiencing repeated or extreme exposure to aversive details of the traumatic events (first responders, police officers).
Note: the DSM-IV-TR also had the following: The person's response involved intense fear, helplessness, or horror. This component was removed from the DSM5 because few combat troops actually admit feeling intense emotions. Some of this can be due to denial but they were too mentally focused on dealing with the threat to feel their emotions until later.

I noticed that this part (the criteria) is written as it is found in several articles on the internet. Unless plagiarism isn't an issue, I would suggest rewriting this part (only minimal changes in sentence structure to avoid plagiarism).

Wide-scale, interpersonal mutilation is horrible. However, war atrocities are so horrific for American because wide-scale, interpersonal mutilation is seen so rarely in America. (Thank

God, thank democracy, thank the National Rifle Association or whoever).

Most accidents don't even mutilate in America…and they're just accidents! People get hurt, not mutilated. That's why we call them accidents. Even if there is horrible mutilation, there's usually no malice or "axis of evil" involved. A train crashes, a plane crashes, a bridge collapses. It's just happened because someone seriously screwed up. They didn't mean to hurt themselves or the others.
They just didn't pay attention, they just got distracted, they took a foolish chance and lost, they zigged when they should have zagged, etc. It happened, it's over in a flash, then everybody goes back to the picnic.

The majority of Americans are still horrified by widespread death and destruction. American's don't do widespread interpersonal violence…not like other countries have and are still doing. I've repeatedly read and heard that America is a violent society. Well excuse me. We've never had a Hitler doing mass murder in the pursuit of the "master race", an Idi Amin slaughter of millions in Uganda or Pol Pot's Cambodian "killing fields".

We are a democracy. Everyone gets an opinion. Nobody is supposed to get killed just for having a different opinion.

It is still un-American to kill women or children. Thank God or whomever.

American's don't do civil wars for a lifestyle. One was enough, apparently. However, the Shiite's and Sunnis have fought since Mohammad died (two thousand years ago), and sub-clans and sub-sub-clans have gone at it for decades, if not millennia. England's War of the Roses lasted thirty years. The Alodarian Empire's 800-year-war lasted…800 years. The only brother-against-brother, father-against-son the Americans do any more are NFL football, NBA playoffs and NASCAR races.

B. Presence of one or more of the following intrusive symptoms associated with the traumatic event(s):

…

(1) recurrent and intrusive distressing recollections of the event, including images, thoughts, or perceptions. **Note:** in young children, repetitive play may occur in which themes or aspects
of the trauma are expressed.

These intrusions can be memories, visual images, smells and auditory distortions. They may be what the civil war veterans having "the reverie" were doing. They were probably lost in intense remembering of the horror of seeing tens of thousands slaughtered in one day on both sides.

Three Civil War generals said about the carnage, "This isn't war, it is murder". D.H. Hill, Evander Law and William T. Sherman.

I have repeatedly heard of another subcomponent of these intrusive recollections I call "**bleed-throughs**". These bleed-throughs aren't the full-blown flashbacks and they aren't the reliving nightmares. They are partial sensory experiences of the past memories overlaid onto the present. They can be visual, auditory, olfactory (smell) or taste. They are mini-flashbacks.

I had one student of mine in a General Psychology class tell me her combat-veterans-sister saw blood on the inside of a taxi-cab in Indiana. It was not really there. It was a bleed-through of a traumatic scene she witnessed in Iraq. She had pulled civilians women and children people out of a taxi car that had been mistakenly machine-gunned by American troops at a checkpoint.

I've heard of returned American Vietnam war combat troops seeing Viet Cong uniforms overlaid on Eurasian civilians back stateside.

Unfortunately, these bleed-throughs also trigger intense feelings that can last for days. The viewer is very confused and emotional for quite a while after misperceiving these things.

(2) Recurrent distressing dreams of the event.

The information I have from 45-years of clinical practice of both severe combat and civilian PTSD shows that these nighttime "flashbacks" (exactly reliving the trauma in a nightmare) are much more frequent than flashbacks in the daytime. I would be open to other practitioner's input or researcher data on this issue. They are more frequent if the person is intoxicated, day or night time.

The only full-fledged, daytime combat flashback I ever heard of was with an intoxicated man. He thought he saw NVA tanks on the streets of his Missouri town. He started shooting at them with his service pistol. Luckily, he didn't shoot or kill anyone.

(3) Dissociative reactions (e.g., flashbacks) in with the individual feels or acts as if the traumatic event(s) were recurring. (Such reactions may occur on a continuum, with the most extreme expressions being a complete loss of awareness of present surroundings.)

This "acting or feeling as if" is pure "transference" in the Freudian term. We act or feel "as if" the past is happening again. However, it is not a sexual surge as Freud analyzed it. It is massive horror, anger, fear and on full combat alert.

(4) Intense or prolonged psychological distress at exposure to internal or external cues that symbolize or resemble an aspect of the traumatic event

There is so much distress now because there was so much distress then. The bizarre part is that they weren't as aware of the stress back then because the majority of their attention was focused on
trying not to get killed!

(5) Marked physiological reactivity on exposure to internal or external cues that symbolize or resemble an aspect of the traumatic event

...

This physiological reactivity component of PTSD is most misunderstood and vastly underestimated aspect of PSTD sequellae.

It's not simply that they are physiologically stirred up to stimuli that symbolize or resemble an aspect of the traumatic event. They often react

as if!!!!

they are going to get hurt again.

We're not talking about being upset. We're not talking "bummin'" We're not talking bad-hair-day. We're talking "fight or flight", "kill or be killed, adrenalin-charged, in-the-firefight-again, I or my comrades are gonna be shot again if I don't attack first (if they have been wounded).

I don't know the body's specific physiological responses to being shot or wounded by any projectile. Part of that depends on what part(s) of the body are hit. Part of that will be the specific projectile involved. I do know there is a massive flood of steroids (both adrenalin, noradrenalin, cortisol and others) to help the body defeat the enemy and to help the body recover from the wound. There is also massive central nervous system arousal (increased sympathetic nervous system, decreased parasympathetic nervous system activity).

Adrenalin is the most commonly known of the stress steroids. During attack or threatened attack, it gets dumped into the blood stream, making your heart pound, your voluntary muscles fed and energized by the flood of sucrose the adrenalin releases, your respiration is hard but slow, you blood thickens (to prevent bleed-out), your skin thickens to prevent bleed-out, your mouth is dry, and all of your involuntary muscles are clamped down. The steroids also help the body heal wounds faster. All parts of your physical body are on a synchronized excited-but-shut-down, ready to go the whole ten yards, do-or-die status.

This steroid-fed mental and physical arousal also makes their central nervous system mentally process things for danger, too.

Guess what? When your body feels like you're in danger, your mind is going to perceive even the innocent or harmless as dangerous. It's called "transference-based misperception". It's called "emotional overlay".

Physical movements towards the "locked and loaded" vet are going to be felt like an attack. They are going to respond to the misperceived attack with a counter-attack.

All sounds (but especially human voices) are possibly going to be misinterpreted as threats. This goes for how they interpret your tone of voice, what you said and how you said it. You are going to have one very paranoid, "locked-and-loaded" person on your hands. They are an instant management issue.

You will often see that a combat vet is really escalated by other people arguing. Why? Because there is a lot of shouting and commotion during combat. It's stimulus generalization, again. What resembles part of the trauma from the past will trigger the emotional response to the traumas again in the present.

There has been research back to post-Vietnam documenting combat veteran's adrenalin hypersensitivity. They physiologically react more to laboratory injections of adrenalin with higher sympathetic reaction (arousal of voluntary nervous system) and, therefore, stronger parasympathetic (shutting down the involuntary nervous system) responses.

This is part of the classical conditioning that occurs in any trauma:
The initial event automatically triggers adrenalin. That rush gets paired with the all parts from all sensory modalities (sound, sight, touch, taste, smell) of the life-threatening event. Then events that are "similar enough" to the initial trauma can trigger other adrenalin responses, even though the person isn't really in another life-threatening situation.

...

The classic combat veteran example is their overreaction to the Fourth of July fireworks. The flashes and, especially, sounds of fireworks are similar enough to the sounds of combat artillery and automatic weapons chatter. Those sounds initially trigger high-alert, fight-or-flight physiological and mental arousal. Every wounded combat vet I knew of was very frazzled and frayed by the end of the Fourth of July holiday.

I find it incredibly ironic and the men and women who risked the most to preserve our independence by putting themselves in harm's way are the ones that suffer the most from it.

I have written letters to major newspapers AND the veterans' department asking for help developing a special "Spare the Vet" program to reduce the stress of illegal firework detonation on this wonderful (for everyone but the combat vet holiday). Nobody has ever taken me seriously enough to even bother to respond to me.

More follows below about the classical conditioning (look for the stick-figure of Pavlov and his dog).

C. Persistent avoidance of stimuli associated with the trauma and numbing
of general responsiveness (not present before the trauma), as indicated
by three (or more) of the following:

(1) Avoidance of or efforts to avoid distressing memories, thoughts or feelings about or closely associated with the traumatic event(s).

It is classic that severely traumatized victims initially try to avoid (suppress) all thoughts and feelings and specific memories about the trauma(s). There can also be some automatic forgetting (repression) of the trauma. Repression has been demonstrated among civilian trauma victims. Don't hassle me about the False Memory Syndrome. I've got a "False Memory Syndrome" lecture subsection in several of my other CEU courses.

The tricky thing about trauma is that it teaches the victim significantly different and erroneous (we used to think, pre-911) perceptions and beliefs about our world and about ourselves.

Within the psychological realm (pre-9/11 attack), children are helpless victims but psychologically healthy adults are potent and powerful. Our parents protect us from abuse and victimization. We protect our kids from the pervert in the bushes God is on our side.

All of those illusions went up in smoke and dust and steel and drywall and body parts on 9/11/2001 for our country. Those illusions are dispelled on a daily basis in every town and county and state regarding physical and sexual abuse.

Regardless of your philosophical take on war (in general or in particular), talking about and examining trauma helps. Most victims won't know that until they try it…for a year (not just two sessions).

Another tricky thing about trauma victims is that they usually try to cope with it on their own before they seek external, professional help. The sad thing about how they try to cope with it on their own (with both civilian and combat PTSD), is that they often go through all of the same bad stages, using ineffective coping or avoidance mechanisms before they hit the wall and seek external help. Those bad stages/ineffective coping mechanisms/ avoidance mechanisms include: taking it out on others, abusing prescription drugs, abusing alcohol, and using illegal drugs, using sex as a tranquilizer, losing your job, losing your marriage, remarrying quickly to someone less functional than your last spouse, doing "antisocial" (criminal) actions.

> (2) Avoidance of or efforts to avoid external reminders (people, places, conversations, activities, objects, situations) that arouse distressing memories, thoughts or feelings about closely associated with the traumatic event(s).

…

This avoidance is because of stimulus generalization. Those places that arouse recollections of the trauma contain stimuli (sights, sounds, smells, tastes, and/or tactile events) that are "similar enough" (stimulus generalization) to those in the trauma that they re-arouse recollections AND intense feelings from the past.

The PTSD victim's best, initial solution to keep from getting re-stimulated is to reduce and/or control all possible stimuli by staying at home as much as possible. This leads to "agoraphobia" (fear of open spaces, especially crowds) in many cases. I have met many Vietnam combat veterans that still live in the woods because they can't handle the overstimulation and triggers in typical civilian cities. I met a Vietnam veteran in a bar recently who had come in for his yearly beer at the local pub. He was walking with a cane, limping along. A mortar round had wounded him in 1968. He didn't know how 1968 was still affecting him until we chatted. He did not even know he had PTSD, 40 years after that mortar got him and damaged his leg forever.

I take it for granted that most people know the classic types of "triggers" but that is apparently not the case. (I will elaborate on the different modalities for triggers later on). However, I'll tell you an example that astounded me when I witnessed how slight and subtle a "similar" stimulus it took to set off a combat vet's alarms.

I was doing psychological testing of an Operation Enduring Freedom veteran who had been wounded with a silver dollar-sized piece of jagged shrapnel in his chin from an IED. One quarter inch closer to his neck and it would have sliced his jugular vein, severed his spinal cord, or both. One other partner of his on the patrol was killed in the blast and several others hurt. His current employer wanted a "fitness to return to duty" evaluation.

His initial wound had happened over a year ago. He told me he thought he was over the combat trauma because he hadn't had any more nightmares for a while.

I tested him in his home. As I talked to him, I looked out his front window over his shoulder and something out in the street caught my eye. My eyes must have quite scanning or my pupils must have constricted or dilated or something. Whatever it was that triggered him, he spun around from his waist up, looking at the direction of my gaze, trying to see what I was looking at. He reacted that strongly just to my fixed gaze. In combat or on a mission, I guess you watch your buddies' gazes, too. They may see something you don't. If anybody misses anything, somebody dies.

D) Negative alterations in cognitions and mood associated with the traumatic event(s), beginning or worsening after the traumatic event(s) occurred, as evidenced by two or more of the following:

1) inability to remember an important aspect of the traumatic event(s).
2) Persistent and exaggerated negative beliefs or experiences about oneself, others, or the world.
3) Persistent, distorted cognitions about the cause or consequences of the traumatic event(s) that lead the individual to blame himself/herself or others.
4) Persistent negative emotional state (fear, horror, anger, guilt or shame).
5) Markedly diminished interest or participation in significant activities.
6) Feelings of detachment or estrangement from others.
7) Persistent inability to experience positive emotions.

In regard to the "inability to remember", the question may really be: Are they unable to recall important aspects or just unwilling? My answer: I've seen both inability and unwillingness. I've seen a lot of unwilling but I've also seen people mentally struggle to get back full memories. However, most of it seems to be unwillingness to remember. Their attitude seems to be: Why should I think about it again? It only hurts more. I don't want to hurt any more so forget it. It's in the past.

...

However, I must comment that I have seen two cases where the soldier totally could not remember major aspects of the trauma. The first involved a Vietnam veteran. He told me that "something horrible happened and he turned and ran toward the mess tent seeking help. Then all of a sudden, he stopped and realized "They can't help me". He stopped running, ambled off and never remembered what that "something horrible" was. The second case involved an Air Force intelligence officer during the Vietnam War. His duty was photo reconnaissance and he was stationed all four years stateside except for three months in Vietnam. He eventually realized that he had little memory of his three months "in country". This amnesia was dramatically brought to his attention during a physical exam he received right after the came back. During that exam his doctors discovered three, six-inches long, parallel healed-over scars on his chest running horizontally. Those scars were not on his chest before he went to Vietnam. He has no recollection of what produced those scars. We are working to try to help him remember their source as I write.

In regard to markedly diminished interest or participation in significant activities: this part contains several aspects. First, the civilian world they return to is usually much less challenging and threatening. Since the civilian world is much less threatening, it is perceived as less important, less noble, less real, etc., by some combat veterans.

Secondly, the release of adrenalin during patrols, combat operations and firefights produces a well-documented high at first and a "coming down" at the end of the adrenalin "dump". Adrenalin can be just addictive as any other mind-altering drug.

Nothing short of high-risk hobbies, high-risk occupations in the civilian world or robbing banks comes even close to the adrenalin "high" of combat (if you don't get shot up). This is why many combat vets take up high-risk hobbies (such as sky-diving) and occupations (police and fire). That doesn't entirely explain bank-robbing.

Third, the question really is diminished interest in "significant" activities" as define by whom? Our American civilian society makes great noise and great money off sports events, elections, politicians and celebrities. The combat vet keeps forever focused on physical survival of him and his loved ones. That is what is "significant" to them and they know (as most others find out) that physical survival is the most important thing on this planet. The combat vet will only be able to let his guard down totally about those in harm's way when they are no longer in jeopardy of dying. Unfortunately, with America's tendency for "perpetual war", that leaves them always on guard and always grieving when they hear of more American military casualties.

The other part of this issue is below. They have feelings of detachment or estrangement from others.

This detachment can be emotional numbing (restricted affect), preoccupation with what is happening to his guys still in the fight, mental preoccupation with what the veterans experienced, himself, and the feelings of many combat vets that the civilians around them.

a) Wouldn't understand what the veteran went through, b) don't care what the veteran went through, or c) don't want to hear about what the veteran experienced. He/she might be correct on all three counts. He/she might not be correct on numbers 2 and 3. He/she is probably correct on number 1 except for other combat vets...and they weren't willing to talk about it at first, either, except since the Vietnam War.

An example of alienation: Did you hear the riddle the one Marine asked the other one as they pulled out their slain comrades floating in the Perfume River during the 1968 battle of Tet? The riddle stemmed from their slain fellow American troops having their testicles and penises cut off and sewn shut in their mouths. The riddle was the following: Do you think they bled to death first or choked to death?

I did not say it was a funny riddle, did I? How often have you heard any Vietnam vet joke about or talk about the atrocities he

...

saw or performed? None, I'll bet. The horrible degradation (beyond comprehension for most) of combat produces the silent secrets nobody else knows.

During that same combat, a friend of mine was on a mission up the Perfume River during the 1968 Tet offensive to sabotage an enemy position. As he and his patrol swept up the river to land, they came upon the severed heads of dozens of Green Beret American troops looking out at them from their supposed "secret" landing spot. You don't see that re-enacted on the Military Channel's Navy Seals or Green Beret episodes. Not even the military will admit that and share these men's grief.

Another example of alienation: Jane Fonda went to North Vietnam in 1972 to protest America's presence. She posed with North Vietnamese anti-aircraft gunners and made anti-American statements on Hanoi radio. She later apologized for those actions that earned her the nickname "Hanoi Jane". However, her actions and statements have never been forgiven by many men who served in combat in the Vietnam War. Her presence on any television or movie still evokes anger and contempt for many Vietnam veterans, no matter how many millions of dollars she has made or how many apologies she gives.

Here's another (nicer) perspective on the combat alienation and distance from civilians after they return. This is circulated on emails on the internet. It was done by Cpt. Allison Crane, RN, MS, Mental Health Nurse Observer-Trainer, 7309th Medical Trainer's Battalion:

When a soldier comes home, he finds it hard....

....to listen to his son whine about being bored.

...

.....to keep a straight face when people complain about potholes.

....to be tolerant of people who complain about the hassle of getting ready for work.

....to be understanding when a co-worker complains about a bad night's sleep.

......to be silent when people pray to God for a new car.

...

....to control his panic when his wife tells him he needs to drive slower.

...to be compassionate when a businessman expresses a fear of flying.

....to keep from laughing when anxious parents say they're afraid to send their kids off to summer camp.

....to keep from ridiculing someone who complains about hot weather.

...

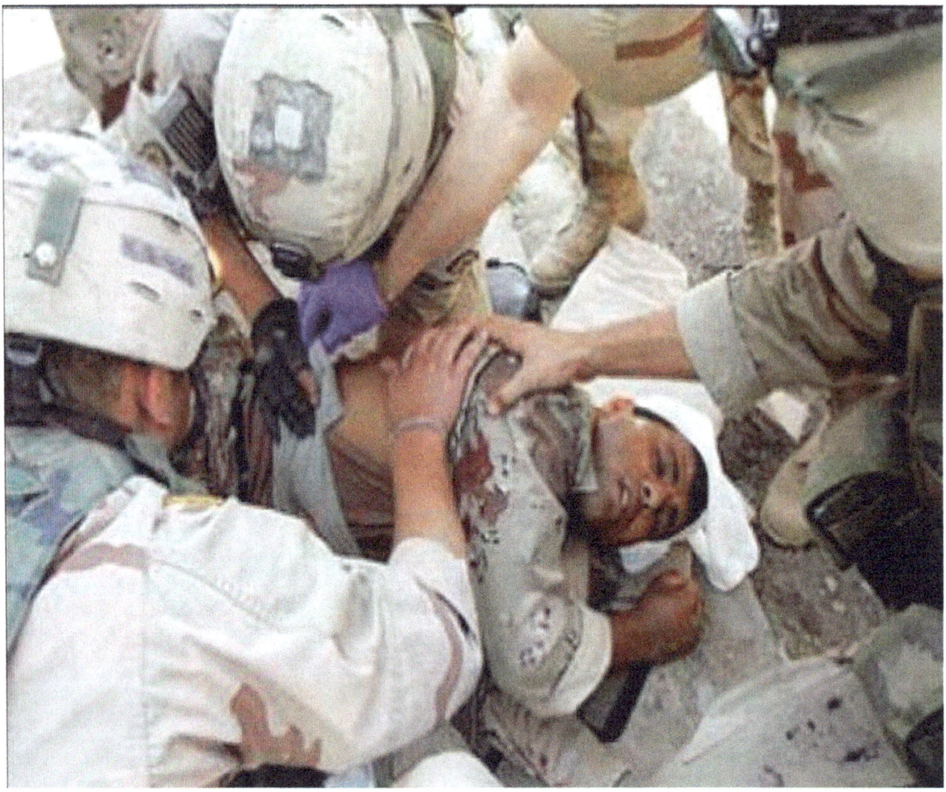

....to control his frustration when a colleague gripes about his coffee being cold.

....to remain calm when his daughter complains about having to walk the dog.

...

...to be civil to people who complain about their jobs.

....to just walk away when someone says they only get two weeks of vacation a year.

...

....to be happy for a friend's new hot tub.

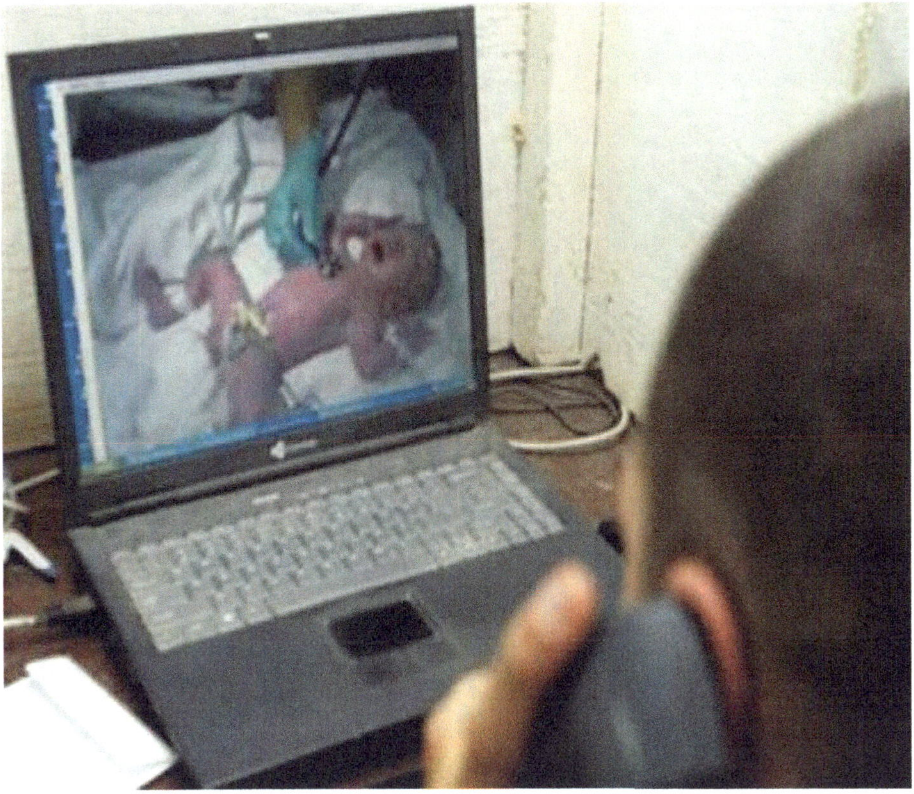

....to be forgiving when someone says how hard it is to have a new baby in the house.

...

....not to punch a wall when someone says we should pull out immediately

The only thing harder than being a Soldier..

...

Is loving one.

Oh, yeah. That goes for their families, too.

How long might their turmoil last?

I saw an example of long-term, war-related emotion at an innocent high school play one year, around 1999. A little country school was putting on a play about the history of America. They did the George Washington Cherry tree skit, a World War I skit, a Depression skit, and then a World War II skit.

For the World War II skit, a young man in the class did a monologue of what life "on the front" was like. The crowd was silent out of respect, but I looked around and saw a couple of white-haired ladies (WWII vintage) crying. I'm sure the young

...

boy's innocent little monologue had triggered painful memories for them.

1999 minus 1945=54 years.

Then I did a Combat PTSD workshop in Seattle, Washington, in 2009. A wife of a combat veteran from Vietnam told me her combat-veteran husband suicided 60 years after combat. He had both legs blown off by a landmine. However, he came back and was a successful businessman without any obvious sequellae (aftereffects). He retired from business and became a pastor without any obvious sequellae. Then a military helicopter crashed in Iraq, killing all 18 soldiers on board. He turned to his wife, said "I can't take it any more" and killed himself.

In 2013 I interviewed a 94-year-old WWII combat medic who was in "for the duration" of the European theater from England (our entry-point to Europe on D-day) to Berlin, Germany (the endgame). He saw lots of tissue damage. He said he was ½ mile from the front, terrified all the time. He still has nightmares of how "butchered" our boys were. He still gets "the shakes" from out of the blue.

1941-2013=72 years!

Once the vet gets over his Emotional Tsunamis, he or she still has to work through the following, no matter how much alienation and sense of entitlement ("You owe me because I did … or suffered …") you feel, a) there are millions of civilians who have their own pains and won't go out of their way for you, b) there are millions of civilians traumatized in other ways (rapes, assaults, 9/11) out there that feel the same "Nobody understands how I feel" as you do, c) nobody owes any of us what we think we're owed, d) you've got to make what you can of the rest of your life with what you've got left (time, body parts, job opportunities, support system), and e)…(fill in your blanks).

Much of this is "working through" is going to involve grieving.

Grieving involves sadness and anger from loss that you can't undo. The loss was unfair, painful, unjust, etc., and you would have done anything to avoid it if you'd only known better. But you didn't know. And you didn't do anything that stopped it. And your wife didn't. And you children didn't. And your parents didn't until it is too late...and you might have to pay the consequences for the rest of your life.

Heavy, man! A life-changing event...for the worse.

As far as restricted range of affect (e.g., unable to have loving feelings), sense of a foreshortened future (e.g., does not expect to have a career, marriage, children, or a normal life span):

Some of this restricted affect involves a generalized diminished range of affect (flattened affect). The returned combat vet does not get too excited about the normal victories and losses of civilian life because few of these events come close to life-or-death matters that are still so fresh in the combat vets mind and heart.

The restricted range of affect also includes the inability to love. This is probably because they have lost someone they loved, to some degree, in battle. This is much more likely in combat troops from the Vietnam war to the present. This is because of the non-national, non-universal conscription that the United States Army has employed to obtain participants for battle since and including the Vietnam war.

In the World Wars and Korean War, everybody went. You were thrown together with people you never met. You banded into groups based on the military's need and you were all "dogfaces", GI's (Government Issue). You were all treated like dirty, expendable socks.

During the Vietnam era, many got drafted and the military developed the recruiting system they called the "buddy system". You got extra money if you signed up with a "buddy" (a friend from high school or work). You went through boot camp

...

together with your "buddy", did AIT (Advanced Individual Training) together and you both shipped out together.

Unfortunately, anybody's horror is magnified when someone you personally know gets killed compared to how you feel if you don't know the deceased, whether that is in civilian life or military life. It's more enormous when you see your-best-friend-since-2nd-grade-and-were-in-Cub Scouts-together-and-played-baseball-together-and-flirted with-their-little-sister get killed.

The same increased familiarity is being caused by deployment of National Guard units to the Gulf Wars who have much pre-war time together, exchange of personal background information (even having family cookouts together), much more shared civilian-life information and other usual civilian-life experiences)

Griefologists know that your grief is worse when the deceased is someone you knew and cared about. The "buddy" system of Vietnam and the "National guard" system of the current (early 2000s) middle-eastern wars make the grief of losses harder, more intense, more prolonged, more painful.

It makes the combat vet reluctant or unwilling to get close to anyone else. They don't want to feel that much pain again. Nothing (including the wonderful soothing of a deep, possibly everlasting love) is worth the pain caused by the unjust, unfair losses they have already experienced.

Their sense of a foreshortened future is based on their direct witnessing of shortened futures of those around them who died. Their hometown buddies, their high school football team members, their fellow trumpet players.

Those unfairly killed now include women and children. The men of WWII witnessed the death of women and children who got caught in the onslaught of counterattack to the Japanese/German/Italian axis. Many of them, to this day, can't stand to hear children cry. Their WWI father's didn't' tell them about that horror because their WWI combat relatives didn't shoot at children. They weren't around. They didn't shoot women.

They weren't anywhere near the trenches. WWI was fought by men against other men, slugging it out "mano- a-mano", hand-to-hand, bayonet against bayonet.

Unfortunately, the battles of WWII, Vietnam and since are fought in the hamlets, rice paddies, streets, alleys and open markets of civilian life. To make matters even worse, from the Vietnam War and since, innocent men, women children in the to-be-liberated country get caught in the crossfire as our soldiers fight the combatant men, women and children of the Viet Cong, Shia and Sunnis, and "insurgents".

Therefore, the combat vet since the 1960s stays more distant from women and children if they have seen women and children combatants or fatalities.

> The "Good Death" fantasy or expectation in America involves the expectation that the average individual (you) is going to die
> a) Quickly,
> b) Painlessly,
> c) While asleep,
> d) Justly,
> e) Peacefully,
> f) At an old age,
> g) Having "done it all" (the "Bucket List")
> h) With their children and grandchildren by their bedside,
> i) With intact body (i.e., not mangled), and,
> j) Going to Heaven (or a Peaceful Other Side).

Wars after WWI showed American soldiers this "Good Death" doesn't happen to everyone. Wars after WWII showed Americans that women and children aren't always warm and fuzzy. They want to kill you sometimes. They try to kill you sometimes. People don't just get shot and die. They get blown to bits, they get mutilated, they get assassinated, they get blown to a bloody pulp.

E. Marked alterations in arousal and reactivity associated with the traumatic event(s), beginning or worsening after the traumatic event(s) occurred, as evidenced by two or more of the following:

(1) Irritable behavior and angry outbursts (with little or no provocation) typically expressed as verbal or physical aggression toward people or objects.
(2) Reckless or self-destructive behavior
(3) Hypervigilance
(4) Exaggerated startle response
(5) Problems with concentration
(6) Sleep disturbances (difficulty falling asleep or staying asleep or restless sleep).

How long can any or all of these trauma sequellae last?

We can also discuss Complex trauma (CPTSD) in war veterans after prolonged and repeated trauma in this section. **https://www.heroesmile.com/complex-ptsd-symptoms-and-treatments-for-veterans/#:~:text=CPTSD%20occurs%20in%20people%2C%20including,war%20or%20other%20life%20stressors**.

I saw a newspaper article about a World War II Battle of the Bulge (very cold, very snowy, very scary) veteran who had a "flashback" when he was caught in a snowstorm (very cold, very snowy, very scary) in Wyoming in 1998. That's about 43 years later.

I've seen many WWII troops well up and cry when they finally talk about their experiences on the history channel programs paying tribute to them.

I worked with a WWII-era combat medic in 2013 who said he still gets the shakes in his arms for no good reason. He said they initially thought it was from Parkinson's disease. He also said he would feel real anxious for no known reason and still has weekly nightmares of "how badly the boys were butchered". He went the duration of the European theater war as a medic. He saw the slaughter from Calais (France) to Berlin (Germany). He said he was sometimes half-a-mile from the front and "scared all the time". He was 94 when I interviewed him, 72 years after seeing his first slaughter.

I still see Vietnam combat vets who are still "irritated" (their words, among other things) by the 4th of July fireworks being shot off around that holiday.
I'm sure many of the combat vets from the Gulf War actions will experience the same irritability or jumpiness about fireworks.

They are actually being re-stimulated (put in a heightened state of physiological and mental arousal) by the fireworks. The strings of firecrackers being set off all at once especially bother them. Why?

...

Because that string of rapid pops and cracks closely imitates the sounds of automatic small-arms fire they experienced during combat. That sound for the combat vet is associated with death, destruction, pain, anger, fear and guilt. For the teenage boy, those sounds are only associated with pretty colors, watermelon and the drink of choice (for them or their parents).

I know many other combat vets that stay inside their house the two weeks before and the two weeks after the 4th of July (because people shoot fireworks off that early and that late around the actual holiday. Many of them drink the entire month and often get into great legal trouble.)

Even severely traumatized civilians carry the scars for decades later. I had a 53-year-old female therapy client who came to me after she had a "flashback" that detailed her father sexually molesting her when she was three years old. That's 50 years later.

There are long-term effects of female, civilian abuse victims as the following study reports: "Child abuse and other traumatic early experiences forever alter a woman's brain chemistry, setting the state for future psychological problems", says a study out Wednesday.

The study in the Journal of the American Medical Association offers the first evidence in humans that early trauma can change the brain's response to stress and raise the risk of mood and anxiety disorders later in life, says lead author Charles Nemeroff of Emory University in Atlanta, Georgia.

Among the findings:

Women who were abused as children were four times more likely than other women to develop excessive stress responses to mild (my underline) stimuli.

Women who were abused and who now have an anxiety disorder or depression are six times more likely than other women to suffer an abnormal stress response."

Springfield (Missouri) News-Leader", August, 2000.

Note that this newspaper article said the women developed excessive stress responses to mild stimuli. How long can a "mild" trauma affect an organism? I have seen laboratory studies on rats that showed they have excessive emotional reactions to a one-time immersion in ice-cold water for a month! I have worked with many World War II veterans who were still emotionally overwhelmed by some "triggers" (current reminders of their combat experiences). I have seen Vietnam vet studies that show they have elevated physiological reactions to simple physical exams 40 years later.

These same long-term results are well documented in combat vets. I can't find the citations, yet. I'll post them later.

...

NON-DSM5 COMBAT VETERANS

PTSD TREATMENT ISSUES

There are three areas that were in the first DSM components of PTSD that were specifically related to Combat PTSD that are not there anymore. I believe they got deleted when PTSD got applied to civilian PTSD. They are 1) Guilts, 2) Anger or Distrust of Authority and 3) Moral Injury. They got put back into the DSM-5. There is one factor that has never been in any version of it: loneliness.

There are many **guilts** specific to combat veterans that are not typically seen in civilian PTSD victims. Those include a) survivor guilt, b) participant guilt, c) nonparticipant guilt, d) atrocity guilt, e) friendly fire guilt and f) guilt about killing another human, be they man, woman or child.

I will discuss those guilts in detail later in the treatment part of this manual.

The Angers

The second frequent issue for combat veterans that isn't is the DSM any more is anger or distrust of authority. Somebody in the 1970s said "Don't trust anyone over 30". There have been many proven unethical treatments of the troops by the United States military and there are more lingering in the abyss of research and/or investigation, unproven but suspected.

The first involved using American soldiers for atomic-bomb guinea pigs. ("Killing Our Own" @ http://www.ratical.org/radiation/KillingOurOwn/KOO3.html, 2009.)
An unknown number of American troops were exposed to radiation from atomic bombs in Nevada testing grounds, starting in 1951. They were not given the current standard of "informed consent" of human research required by all present-day medical and psychological research subjects.

The survivors and their families were compensated for their injuries sometime in the 1990s but I cannot find information sites for this.

American soldiers in Vietnam suffered from a second unethical lapse by the Department of Defense. This lapse involved Agent Orange. See the VA's information site (www1.va.gov/agentorange) for all information.

Agent Orange is a defoliant. That means it strips plants of leaves. It was sprayed in Southeast Asia to strip the jungle of its leaves so that the Viet Cong and NVA has no shelter. It worked very effectively. It contained one of the most toxic chemicals known to man: dioxin.

Dioxin is so powerful that one teaspoon full will kill a thousand humans. One of the manufacturers of it (Dow and Monsanto chemical companies) sent a memorandum to the Pentagon warning them against using it near humans. The pentagon ignored that warning and sprayed it without consulting the commanders on the ground in those areas. It rained down on our troops in the jungle on patrol. They breathed it, they had it fall on their skin, in their eyes and mouth. Soon after the vets returned, many of them developed strange skin cysts, other soft tissue disorders (including cancers), and many blood disorders.

The troops were not only exposed to it from the air. A friend of mine who was around the Phu Bai air complex in Vietnam said they used to <u>cut in half</u> the empty 55-gallon drums housing the Agent Orange and <u>made them into barbeques</u>. They <u>cut the tops out of other drums</u> and filled them with sand to make bunkers for protection against artillery and mortars. They cut other drums and half and used them for the latrine buckets. They cut them in half and used them for showers. They were breathing it on a daily basis and washing themselves in it.

The DOD's initial research showed "no relationship" between exposure to Agent Orange and the veterans' illnesses. Unfortunately, their initial research "exposure group" was the Air Force personnel loading the sealed drums onto the spray

...

airplanes. Those "exposure group" personnel did not actually touch, breathe or eat any of the Agent Orange. The government/chemical company settlement established a multi-billion dollar fund for in-country Vietnam veterans and their children (since their children were found to have higher incidence of birth disorders and neurological defects, too).

The Gulf War, brief as it was, brought its own health and psychological issue, called The Gulf War Syndrome. Symptoms attributed to this syndrome have been wide-ranging, including chronic fatigue, loss of muscle control, headaches, dizziness and loss of balance, memory problems, muscle and joint pain, indigestion, skin problems, shortness of breath, and even insulin resistance. Brain cancer deaths, amyotrophic lateral sclerosis (also known as Lou Gehrig's disease) and fibromyalgia are now recognized by the Defense and Veterans Affairs departments as potentially connected to service during the Gulf War, possibly due to ionizing radiation (50 FR 34459, Aug. 26, 1985, as amended at 54 FR 42803, Oct. 18, 1989).

The newest areas of substantiated anger include the burn pits in Iraq and Afghanistan, the pulmonary effects of the Iraq oil well fires, the Camp Lejeune (and many other military base) tap water and chemical contamination and "The Suicide Pill/Kill Pill", Lariam/Mefloquine.

The first issue involves the chemical contamination of veterans' lungs back to Vietnam by exposure to the toxic fumes generated by the massive trash dumps in Vietnam and after that. The practice gained great exposure recently in Iraq and Afghanistan when it was found that a subsidiary of a company owned by ex-Vice President Richard Cheney (KBR) had a contract to install, run and maintain massive incinerators throughout those countries to reduce exposure to toxic fumes from trash burning. However, K. Kennedy wrote on 2/2/15 that many of those incinerators were not installed and most of those that were installed were not repaired if they broke down. She also wrote that the United States Supreme court ruled that American soldiers could sue KBR in a class action lawsuit. There is a VA website with the address of:

(**https://veteran.mobilehealth.va.gov/AHBurnPitRegistry/#page/home**) for those to register for burn pit-related health problems. If that address does not get you there, just type in "VA burn pit registry". That will get you to the site. She said her data over 13 years on in-theater veteran's health problems showed "Chronic obstructive pulmonary disease increased steadily from a rate of 98 cases per 10,000 in 2001 to 218 per 10,000 in 2009, before going back down to 147 in 2013. About 92 percent of COPD cases are diagnosed in people older than 45, according to the American Lung Association—not in healthy service members, only 9 percent of whom are older than 41. Chronic sinusitis increased from a rate of 71 in 2001, to 245 in 2009. Cardiovascular symptoms increased from 224 in 2001 to 280 in 2009. Tumors increased from 91 in 2001 to 2005 in 2009. Symptoms of neurologic conditions increased from 70 in 2001 to 252 in 2009."

There is also a Gulf War oil well registry. To schedule an exam for possible lung damage due to exposure to those fires go to **http://www.publichealth.va.gov/exposures/gulfwar/** or call 800-PGW-VETS (800-749-8387). The American government and American military have not been as evasive and in denial about the possible effects of exposure to the well fires as KBR has been about the burn pits.

The third issue many military members, specifically Marines, are concerned and angry about is the contamination of the grounds and drinking water at Camp Lejeune, the training base for new Marine recruits, and other military bases. Various toxins were dumped around the base from 1953 to 1987 and contaminated the drinking water with the "VOCs (volatile organic compounds) known as PCE (Tetrachloroethylene aka Perchloroethylene), TCE (Trichloroethylene), DCE (Dichloroethylene), Vinyl Chloride and BTEX (Benzene, Toluene, Ethylbenzene, and Xylene).These chemicals are either known or suspected human carcinogens. Many Marines, Sailors, their families and loyal civilian employees have been affected by the contamination in various ways including, but not limited to: liver cancer, kidney cancer, breast cancer, bladder cancer, ovarian cancer, prostate cancer, cervical cancer, lung cancer, leukemia, non Hodgkins

...

lymphoma, liver disease, miscarriages, birth defects (cleft palate, heart defects, Choanal atresia, neural tube defects, low birth weight, and small for gestational age" (The Few The Proud The Forgotten @http://www.tftptf.com/). The registration site for this issue is at **http://www.tftptf.com/63512/index.html** or through the VA. However, I suggest you submit your information directly to the VA to avoid any loss of information transfer but I cannot find a specific referral mechanism to do so.

Heightened anger seems to be major emotional change for combat veterans, initially, and seems to be the major emotional difference between combat trauma and civilian trauma. Besides the social inequities mentioned above, I also believe heightened anger is so prevalent because of the interpersonal differences between civilian and military traumas. In civilian traumas, the victim usually reacts by "fight or flight", fight them off or run away. However, in combat, the soldier is in a "kill or be killed" situation and all other options are off the table once the order for "guns up" is issued. The combat troop is ordered to open up fire and don't stop until the "cease fire" order is given. The combat veteran is trained and has reacted with "kill or be killed" on an ever-mission status. That is why the combat vet has so much anger and irritability when he/she returns stateside.

The last issue is what veterans call "The Suicide Pill" or the "Kill Pill". It is their nickname for the anti-malaria drug the DOD gave them before they went to many different areas of the world. The generic name for that drug is mefloquine. It was manufactured by Hoffman-Laroche in America until 2009. It is still available in America under different generic names. Some of the trades name are Lariam, Mefloquine or Mefliam. The DoD recommended the drug not be prescribed for service members after several incidents of irrational homicidal rage occurred in highly trained combat veterans both back home and abroad. The most extreme cases where Lariam was possibly involved include four soldiers from one unit who murdered their spouses at Ft. Bragg and SSgt. Bales who sauntered off his post in Afghanistan and murdered 17 Afghanistan civilians in March of 2012. The following is a report Roche filed as required with the FDA:

"Homicide [REDACTED] A patient of unknown demographics started on mefloquine (therapy details unspecified) for an unknown indication. After an unspecified duration, the patient who was a soldier experienced homicidal behavior which led to homicidal killing of 17 [REDACTED]. It was reported that the patient was suffering from traumatic brain injury (TBI) and was administered mefloquine against military rule (mefloquine is directly contraindicated in patients with TBI as per [REDACTED] rule).

While not named specifically in this FDA report, SSgt. Bales was known to have PTSD and a TBI. The DoD has not affirmed to my knowledge if SSgt. Bales had been given Lariam. Search the internet for videos and printed research by Dr. Remington Nevin. His videos are extremely technical but very detailed and the how's, why's and where's of mefloquine neurotoxicity. (Unfortunately, SSgt. Bales was also known to be sleep deprived, drinking alcohol, abusing steroids and taking some psychotropics. All of these will also make a person very emotionally unstable).

While all medications have negative side-effects, the sheer insanity of the behaviors and actions of many soldiers who had taken it have been documented. It was also noted that many tour guides in Africa were familiar with the "melt-down" effects of the pill long before the military used it. I do not see any registry site for this concern but one veteran I worked with in April of 2015 said he had registered with the VA so contact them if you have a concern in this area. There are many informational articles available on the Internet when you search "Mefloquine", "Lariam", "kill pill" and other subsets. The DoD has not issued any additional bulletins about it to my knowledge since ordering its discontinuation in about 2009. The most current (2018) VA policy statement about Lariam is at https://www.publichealth.va.gov/exposures/mefloquine-lariam.asp.

To summarize the neurotoxic effects of mefloquine, the drug affects the amygdala (rage/fear center), hippocampus (short-term memory/learning center), brain stem (sleep and balance). So the

...

veteran can have all of these symptoms without having been psychologically traumatized or head injured by classic, external TBIs.

Moral Injury (Where was God that day?): the term moral injury surfaced around 2012 to describe the loss of moral compass that many combat vets experience during and after combat. The terms "participant guilt" and "atrocity guilt" were used for the same idea during the Vietnam War. I hypothesize that loss of moral compass was behind much of the heavy alcohol consumption by actual combat veterans after many wars. I also hypothesize that it is related to the many combat vets who took long motorcycle rides and post-war combat vet hitchhikers that came back from Vietnam and other wars. They had to clear their heads. Clear their heads of what?

Clear their heads and hearts of the moral anguish and guilt they had from killing men, women and children and witnessing the stark, brutal butchery it involved. As I said very early in this writing, American men are not accustomed to killing. They are not accustomed to killing other men. They are certainly not accustomed to killing women and children or even killing men that look like children.

A Vietnam War combat veteran wrote a poem to express his moral numbness upon seeing a bloated enemy solider floating in an inlet after a firefight. It was being eaten by an alligator, brains and all. He allowed me to share it. I cannot mention his name due to clinical confidentiality. He gave me permission to use it. His poem is below:

CONVERSATIONS WITH A CROCODILE (1969)

WHAT ARE YOU DOING HERE?

C) I LIVE HERE-WHAT ARE YOU DOING HERE?

I'M HERE TO SET THESE PEOPLE FREE-WHY ARE YOU EATING THAT MAN?

C) BECAUSE I'M HUNGRY

DON'T YOU KNOW THAT THIS MAN IS DEAD?

C) I DIDN'T KILL HIM

I DIDN'T KILL HIM EITHER

C) MAN KILLED HIM-YOU ARE A MAN

CAN'T YOU FIND SOMETHING ELSE TO EAT?

C) IF I EAT THEN I HAVE TO KILL BUT HERE MAN HAS KILLED AND I EAT

WHY CAN'T YOU LEAVE THE DEAD IN PEACE?

C) WHY CAN'T YOU LIVE IN PEACE?

THAT'S JUST THE WAY OF IT

C) MY PEOPLE ONLY KILL TO EAT AND THAT'S THE WAY OF IT-MAN KILLS FOR THE SAKE OF KILLING

WHY DO WE LIVE IN FEAR OF YOU?

C) WHY DO WE LIVE IN FEAR OF YOU?

JUST THE WAY IT IS-CAN YOU NOT LEAVE THIS MAN ALONE?

C) THIS IS NO LONGER A MAN

DO YOU HAVE ANY COMPASSION?

C) IF MAN HAD ANY COMPASSION THEN I WOULDN'T BE HERE--THIS IS NOT MY DOING

I BEG OF YOU TO LEAVE THIS MAN ALONE

...

C) IF I DON'T-WOULD YOU KILL ME ALSO?

SURELY ONE OF US WILL!

C) I AM THE FATHER OF MANY CHILDREN. WOULD YOU KILL THEM ALSO? AND THEIR CHILDREN UNTIL THERE ARE NONE OF US LEFT? IS THIS NOT ALSO THE WAY OF MAN?

THERE ARE THOSE THAT WOULD-WHY DO YOU TROUBLE ME EVEN TODAY?

C) I HAVE NOT TROUBLED YOU-IT IS YOU WHO WILL NOT LEAVE ME IN PEACE

WHY DO YOU NOT LEAVE ME IN PEACE?

C) IT IS YOU WHO DID NOT COME IN PEACE

LEAVE US WHILE YOU STILL HAVE LIFE

C) TAKE THAT WHICH YOU HAVE KILLED WITH YOU OR I SHALL RETURN.

SLOWLY WALKS INTO THE BUSH AT THE RIVER'S EDGE AND ONCE MORE STATES THAT HE HAS NOT TROUBLED ME

While this veteran wrote this poem in therapy to try to deal with his emotional numbness about seeing a bloated man being eaten by an animal, he verbalized to me in our session that he was mostly shocked that neither he nor the two other soldiers in his boat could say anything about the interaction. They sat there in total silence in the boat, neither talking to each other, to the alligator, or to the cosmos. They just turned the boat around and paddled back to their base. They did not say anything about retrieving the dead enemy combatant. They did not do anything to give him a dignified ending. They were all morally stunned and directionless.

I also hypothesize that most of the combat vet suicides after they came back stateside were related to their guilt about violating their and the world's basic morality.

I said early in this book that I have not included the worst of the atrocities the American soldiers have told me they witnessed. Those involve what the enemy soldiers did to their own women whom they believed or knew had fraternized with the enemy. They tell me these things with great guilt and shame because they knew that if we had not been in their country, those women would not have suffered the brutal degradation inflicted upon them. One of them never talked to me after telling me what he witnessed.

"There are no atheists in the foxhole"-attributed to Dwight D. Eisenhower, President, General of the United State Army.

"Some men find God in the trenches; some men lose God in the trenches"-Fred Nolen, 2014.

Loneliness: there has been recently published research on the dangers of loneliness (ex.: **https://www.psychologytoday.com/articles/200307/the-dangers-loneliness**). However, the psychoanalyst Freida-Fromm Reichman started writing about it theoretically since 1959 when she wrote "On Loneliness". I have not seen any research on it regarding combat veterans but I suspect it is very painful for some veterans due to the following statistics:

In World War II 37% of all able-bodied men were in the service. The rest were home doing something for the war effort.
In Vietnam, only 37% of able-bodied men were in the service due to deferments.
In OEF/OIF, less than 8% were in the service.
In all of the conflicts, a much smaller %-age was actually in battle. Therefore, the "Band of Brothers" got smaller and smaller with each passing era. The cohort group of those who really understood, got smaller and smaller. The isolation and loneliness got bigger and bigger. While one may say social medium and the Internet could make up for that shrinking %-age, I did want to

...

point out that loneliness is a major emotional strain for combat veterans and they are the ones who are at highest risk for PTSD, suicides and substance abuse.

Potential Causes of Gulf War Syndrome

Many of our American soldiers in the brief conflict designated "Desert Storm" (2 August 1990 – 28 February 1991), soon became ill with a variety of physical illnesses. After lengthy and intense research, the DOD identified the following factors as potentially related to their illnesses. See the VA.gov website for current information about these illnesses and the VA's accepted responsibilities. Obviously, the combinations of factors differ with individuals, hence it is likely that there is not one single explanation of the whole spectrum of symptoms. However, the following main categories are candidates for causal relationships with illnesses reported by veterans:

- Administration of three vaccines intended as protection against nerve and biological warfare agents. These were:

1. Pyridostigmine, normally prescribed for myasthenia gravis and known to have serious side effects, especially when the person taking it is exposed to heat. It is also known that exposure to pesticides and insecticides (Baygon, Diazinon and Sevin) should be avoided when taking pyridostigmine because they can accentuate its toxicity. Some women who took this drug during pregnancy and have breast-fed infants have seen side effects in their child.

2. Botulinum Pentavalent, an unproven vaccine intended to counteract botulism. It is unlicensed in the United States.

3. Anthrax, to protect against the disease anthrax. This was apparently selectively administered to troops

during the war, and women receiving it were warned not to have children for three or four years.

4. Depleted uranium was used for the first time in this war. It was incorporated into tank armor, missile and aircraft counterweights and navigational devices, and in tank, anti-aircraft and anti-personnel artillery. The scientific information on this deadly chemical has been reported in "Radium Osteitis With Osteogenic Sarcoma: The Chronology and Natural History of Fatal Cases" by Dr. William D. Sharpe, Bulletin of the New York Academy of Medicine, Vol. 47, No. 9 (September 1971). There was no excuse for this human experimentation because the effects of this exposure were known.

5. Smoke and chemical pollutants released by the continuous oil- well fires. Levels of soot, carbon monoxide and ozone have been studied by an Environmental Protection Agency Task Force. The National Toxics Campaign, Boston, Massachusetts, found five different toxic hydrocarbon products in the smoke (1,4-dichlorobenzine, 1,2-dichlorobenzene, diethyl phthalate, dimethyl phthalate and naphthalene), any one of which could induce serious health effects.

6. Old World leishmaniasis, a parasitic disease transmitted by the bite of many species of sand fly indigenous to the region. Non-indigenous people who enter an infected area are known to be more seriously affected by this parasite than the inhabitants. If left undiagnosed, and therefore untreated, it can be fatal. Diagnosis requires bone and spleen biopsy, and the disease can have a three-year incubation period without causing symptoms. It can be transmitted by blood transfusion, and transmitted by a woman to her unborn child. Leishmaniasis was reported as widespread in Iraq and Saudi Arabia. This disease is thought to be responsible for the Pentagon ban, November 1991, against blood donations from Gulf War

...

veterans. This ban was lifted, for unknown reasons, on January 11, 1993.

7. Pesticides and insecticides were used extensively throughout the war to protect against pestilence. It is known that large quantities of DDT, malathion, fenitrorthion, propuxur, deltamethrin and permethrin were used. They are all toxic nerve agents, and many are suspected carcinogens and mutagens.

8. Destruction by allies of Iraqi chemical, nerve and biological warfare weapons resulting in widespread distribution of these toxins in the environment. This problem has now been, at least in part, documented by the U.S. Department of Defense. They are focusing on this potential cause as if it were the only candidate cause.

9. The electromagnetic environment which permeated the battlefield during the war. Veterans were exposed to a broad spectrum of electromagnetic radiation created by electricity generated to support the high-tech instruments, thousands of radios and radar devices in use. This intense electromagnetic field causes both thermal and non-thermal effects, and potentially interacts with the other hazardous exposures and stresses of the battlefield. Electromagnetic radiation can alter the production of hormones (neurotransmitters), interact with cell membranes, increase calcium ion flow, stimulate protein kinase in lymphocytes, suppress the immune system, affect melatonin production required to control the "body clock," and cause changes in the blood-brain barrier.

(from Gulf War Syndrome, Depleted Uranium and the Dangers of Low-Level Radiation, by Dr. Rosalie Bertel @http://www.ccnr.org/bertell_book.html (2009).

However, in 2010, the Veterans Administration announced the following bulletin:

Infectious Diseases Associated with Gulf, Iraq, and Afghanistan Conflicts

On March 18, 2010, VA published a proposed regulation* that will establish nine specific infectious diseases as associated with military service in Southwest Asia during the Gulf War from 1990 to the present and in Afghanistan on or after September 19, 2001. The nine diseases are:

Brucellosis

Campylobacter jejuni

Coxiella burnetii (Q fever)

Malaria

Mycobacterium tuberculosis

Nontyphoid Salmonella

Shigella

Visceral leishmaniasis

West Nile virus

*Consult your VA case manager for details, an updated list and application process.

EPIDEMIOLOGY OF COMBAT-INDUCED PTSD

The current estimates vary widely on the epidemiology (prevalence) of combat PTSD. Dr. Charles W. Hoge, one of the researchers at the Walter Reed Army Institute of Research, reported that one out of eight combat troops were showing

...

symptoms of PTSD. This was published in 2004 in the New England Journal of Medicine.

A long-term perspective on it was given by the results of a more recent study (title and author?) that found 30% of Vietnam veterans had PTSD after about 30 years.

A more recent study by the VA or DOD stated 50% of OIF/OEF vets have it. Their DOD latest figure is 12-20%, but that is diagnosing with a conflicting interest of saving money against benefits payouts (see later segments in this book on that).

DIFFERENTIAL DIAGNOSIS IN ADULTS

The only significant differentiation I want to make is between PTSD (309.81) and Acute Stress Disorder (308.3) is that Acute Stress Disorders tend to occur from natural disasters (tornados, storms, earthquakes, fires, etc). The emotional damage from them will be worse if there is an intentional human involvement (such as if someone intentionally sets the fire that burns your house down).

Natural Disasters vs. Manmade Assaults: 1. rapes, 2. sexual abuse (children), 3. physical abuse, 4. combat)

There are some similarities of the destructive aftereffects across these different sources of trauma, BUT there are many specific differences in diagnosis, prognosis, and treatment. Let's look at each one in turn.

Natural Disasters-floods, tornados, earthquakes, car wrecks, and other natural disasters are more prone to creating Acute Stress Disorders (308.3) (Source: Friedman, 2001). This diagnosis shares many of the same symptoms as PTSD but lasts a maximum of four weeks post-trauma (according to the DSM-IV).

Natural Disasters produce less intense, shorter duration trauma for two reasons. First, they are not intentional, manmade traumas. Second, there are many warning signs of impending serious natural disasters (cloudy skies, winds pick up, minor tremors that tell you a fault-line is near, etc). Therefore, natural disasters do not create nearly as strong emotional damage as man-made injuries. Third, there are many possibilities for **"extinction"** of the traumatic pairing (see more about extinction below). Skies darken over with clouds all of the time but don't produce tornados every time. Winds pick up in any season but don't always produce tornados. Minor earth tremors occur frequently (even around major fault lines) but don't often mean major quakes are going to happen. Fires burn everywhere (in your neighbor's burn barrel, someone's house, someone's pasture) but the fire rarely engulfs everything in your neighborhood. It rains a lot without it becoming a serious flood These are all produce extinction for the serious emotions tied to major natural disasters.

The Classical Conditioning of Trauma

Most of the sequellae of trauma can be directly explained by the Classical (Pavlovian) Conditioning paradigm. Ivan Pavlov, a Russian physiologist, developed the concept working with dog's salivation (to research digestion). The dogs were immobilized by the contraption illustrated below. Note that the dog was immobilized and the procedure done to it without the animal's voluntary control.

...

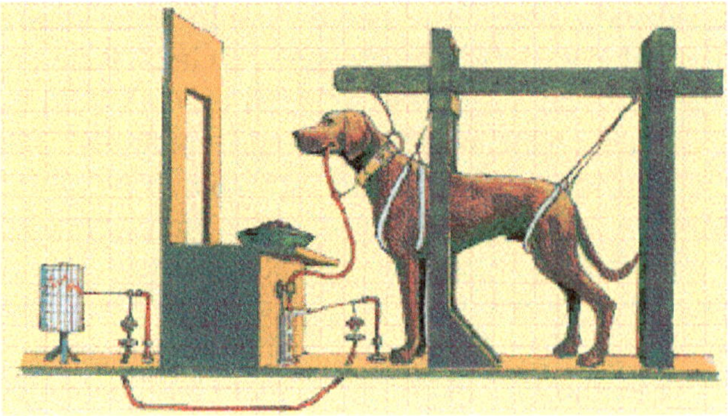

From:
https://images.search.yahoo.com/search/images?p=ivan+pavlov+and+dog+pictures&fr=mcafee&imgurl=http%3A%2F%2F3.bp.blogspot.com%2F_VtUZ-A1VkcA%2FS40vAo8dPhI%2FAAAAAAAAAAk%2FHpbdkmn0XWY%2Fs320%2FPavlov3.gif#id=11&iurl=http%3A%2F%2F3.bp.blogspot.com%2F_VtUZ-A1VkcA%2FS40vAo8dPhI%2FAAAAAAAAAAk%2FHpbdkmn0XWY%2Fs320%2FPavlov3.gif&action=click.

Classical conditioning is learning through pairing. Classical conditioning is learning without choice. It is learning without effort. It can be learning without consciousness.

Ivan Pavlov (1849-1936) studied how saliva helped digest food. He had to deprive (semi-starve) his dogs to make them salivate (to chunks of meat) when he needed their saliva.

One day, he noticed that the dogs salivated when a certain, specific laboratory assistant entered the room. They did not salivate to any other assistant or to Pavlov. He asked the assistant if he could explain their salivation to him. He could not or would not provide an explanation and initially denied he had been slipping them chunks of meat. He eventually admitted he had been slipping them chunks of meat behind Pavlov's back (in violation of research protocol).
The dogs associated (paired) the assistant to the chunks of meat. They salivated to the assistant as if he were the meat.

In Pavlov's classical conditioning paradigm, learning is being produced by pairing a bell ring to meat powder. The animal initially doesn't salivate to the bell ringing. With enough pairing (ring the bell, then give them some meat), Classical Conditioning (learning) will occurred when the bell, alone, elicits salivation.

You can notice your classical conditioning next time you salivate when you smell good food. It's the same process. The smell is paired with the taste which is paired with the past response of food in your mouth: salivation!.

According to Pavlov, ALL emotions, good and bad, get learned through pairing or association with events and that event's stimuli.

When we are children, we feel positive about someone when they treat us nicely, give us food, hugs and praise, and other pleasant stimuli. We associate_that individual with positive, nice things, so we like or love them.

Emotional trauma (negative feelings) gets Classically Conditioned in the same way. A traumatic event happens, then all of the stimuli associated with that event can "trigger" almost-equivalent emotional AND physical reactions. (Some samples of different sensory triggers are listed below in the treatment protocols.) Intense trauma that only happens once can produce conditioning for a lifetime.

In the combat wounded or traumatized, current sights and sounds and smells get paired with the pain and fears and angers of battle and trigger near-identical emotional and physiological reactions without the veterans usually knowing it at first. They are in "full-metal-jacket" mentally and physiologically without realizing they off this planet.

Extinction of reactions occurs simply when the bell (in the above example) rings but no more meat powder comes. Emotional reactions extinguish from most natural disasters

...

simply when events similar to the natural disaster (strong winds or dark clouds, for example) occur again but no tornado hits.

Most emotional traumas from natural disasters extinguish more quickly than manmade PTSD because there are more frequent occurrences of the similar events without the recurrence of the disaster. The skies cloud up a lot more times without a tornado tearing up your house. It becomes windy a lot more times without the tornado coming down and snatching your truck two miles down the road.

Note that I said, "more quickly". Some people suffer extended, true PTSD from natural disasters. The amount and duration of PTSD is always easily calculable, be it from natural or man-made events.

> I believe the most accurate predictor of severity and duration of PTSD, regardless of the cause, is always the **amount of known tissue damage** suffered whether
>
> a) received to your own body,
> b) witnessed to your comrades or
> c) inflicted on others *that you see*!

Can I make it any plainer? Do the math and you'll see! Those who get wounded have the enormous physiological (physical) reaction of the projectile tearing through their body as well as the physiological reactions due to the intense sights and sounds and smells they are experiencing.

No matter how bad everything else is, the unwounded don't have that massive physical reaction to cope with. (Note: that is not meant to minimize the non-wounded's trauma and suffering. It's just a mathematical, physiological reality).

However, I hypothesize the amount of PTSD from received tissue damage causes the anxiety-based symptoms and the PTSD from tissue damage inflicted on others, especially "innocents", when directly observed feeds more into the guilt and grief components. I also have data from my own research with all the limitations and shortcomings of it that shows the physically wounded veteran is rare among the combat vets suicides after discharge to date.

Natural disasters that don't involve tissue damage produce shorter, milder PTSD or mere Acute Stress Disorder. Much of that seems to be due to the "accidental", "It's not personal" nature of the injuries.

Rape-is sexual violence, be it by stranger, date or spouse. Tissue damage can be great; emotional trauma is commensurately great because it's an intentional interpersonal assault. People who love you, like you, marry you, date you, buy you "stuff", whisper sweet nothings" in your ear <u>aren't supposed to</u> brutally assault you. They violated their word and your expectation of them given from their words and initially-kind actions.

...

TREAMENT AND TREATMENT ISSUES

GENERAL THERAPY PROCEDURES

Framing

Effective treatment for any client/patient initially demands some general framework. (I say "demands" and underline it because it is crucial to set up this initial framework. If you don't discuss them at first but later hit one of these issues, it is near-impossible to backtrack and salvage the treatment.)

The general framing for any client/patient minimally involves:

1) Signing of all consent forms and patient rights forms
2) Be clear about "duty to warn" regarding harm to self or others.
3) Establishing mutually agreed upon financial aspects
4) Defining boundaries of client and patient regarding touch, phone calls, charges for phone calls, appropriate hours of phone calls, after-hours
5) crises, calls or visits to the therapist's home, etc.
6) Discussing the positive and negative aspects of therapy (such as negative side effects).
7) Discussing negative "transference" in terms of I, their therapist, may hit their "buttons".
8) Discuss "I never promised you a rose garden". This is the notion that therapy cannot make everything wonderful, forever.

Trauma Therapy Framing: You, the therapist, must be aware that your attempt to do therapy challenges one of the primary symptoms of PTSD listed in the DSM5 (section C, item 1): efforts to avoid thoughts, feelings, or conversations associated with the trauma. You must let the combat vet know immediately that doing effective talking therapy involves invading that defense mechanism (avoidance).

So how are you going to convince the combat veteran that approaching his/her experience is going to be "worth it" or produce any different effect than he/she has experienced in the past when they thought or talked about it: BAD!

You, the therapist, must know and be able to tell them that the "working thorough" of talking therapy involves

a) Sharing their pain,
b) Letting them know the therapist cares about the vet's pain,
c) Letting the combat veteran know (by behaviors and words) that the therapist does not condemn or negatively judge the veteran for their actions,
d) Helping the combat veteran see their actions in a different light (reframing),
e) Letting the combat veteran know they are not alone in their strange (i.e., different) thoughts and perceptions and actions afterward,
f) Giving them help with their guilt (if any), and
g) processing other "erroneous beliefs" they formed from their experiences those persistent, exaggerated negative beliefs and expectations mentioned in section D.2 of the PTSD criteria.

The only phrases the <u>non-combat veteran</u> therapist needs to avoid is the phrase, "I understand" or "I know how you felt". Combat veterans, in particular, are notorious for immediately responding, "No, you don't understand because you weren't there".

1) Discuss **the First Dictum** of trauma therapy:

   Getting Better Does Not Mean Feeling Better…at First!!!!

2) Discuss **the Second Dictum** of trauma therapy: You're not defective, you're emotionally wounded, traumatized. So many people with PTSD caused by attacks by other people have difficulty separating the feelings they have from the abuse from the feelings they have about themselves after the abuse. The classic example of this failure to separate is the rape victim who immediately goes home and takes a shower

…

because they feel "dirty". They can't initially separate the degradation done to them by the rape from their feelings about their own physical body.

3) Discuss **the Third Dictum**: that the trauma victim may also be a victimizer. I recommend the therapist immediately warn the
combat veteran that there will probably be two parts to their therapy.

Part A of therapy is dealing with the traumas they witnessed or received. Part B might be their inability to control their own emotions, especially anger, after they come back to the safe world.
As we in trauma therapy know, Abused people often become abusive to other people.

General Combat Veteran Therapy Issues:

I can think of the ten following treatment issues for combat PTSD. I'm sure I (and you) can think of more as we go along. This list is not mean to be all-inclusive. Listen to each vet-he or she will tell you his/her important ones. Proceed from there.

The first and most important is Stimulus Generalization:
It is crucial to help them identify their "**triggers**".

The "triggers" can be the classic ones (backfires, Fourth of July fireworks, nightmares) and the unexpected ones I will list below and you may learn more. Some of the unexpected I have seen included close thunderbolts, tree lines, helicopters, leaves coming out in the springtime, children of certain ages, any child crying, the clatter of dropped skateboards and present-day military operations (Desert Storm, Afghanistan, Operation Enduring Freedom). The middle-eastern operations will have desert-type triggers that the Vietnam and northern theater WWII guys didn't have.

These triggers can provoke massive emotional floods that can go on for months and be triggered decades after actual combat. I call vets in this prolonged emotional surge "in Full Metal Jacket" (after Stanley Kubrick's movie of the same name).

The triggered vet back in the states is on full combat alert, like he was in harms' way **but** the physiological arousal is in a completely different context (setting) but the vet is almost unconsciously in the old kill-or-be-killed mode and can only be pissed off at everything and everyone.

This is why they are mad all the time when they come back. They're hypersensitive to stress (even benign things like crowds and "normal" conflicts) and get set into "Full Metal Jacket" a lot.

The "high impact" groups (included Vietnam War-related popular films) triggered emotional responses to a detrimental level, sometimes. They were so stimulated that they could not effectively process their emotions during the 1½-2 hour group. They had to rely on each other and the prison psychologist during the rest of the week. I'm not sure if all of them did effectively process their issues. (Also see my comments about medications in the appropriate section, below).
The sounds of helicopters are triggers for Vietnam veterans and ever after because the helicopters were used as the transport mechanism into combat, out of combat and as Medevac's for the dead and wounded.

They were also used for transport, fire support, and interrogation of the enemy. One Vietnam vet told me that US interrogators would take off with a group of enemy prisoners and ask one of them a question.
If he didn't answer, they'd throw him out of the helicopter to "motivate" the other prisoners to answer the question. It usually worked. One soldier I talked to said he'd heard the prisoners scream and hit the ground as he observed from below.

Helicopters don't serve as triggers for World War II and earlier at all because there were no helicopters during those wars. They don't trigger many Korean War vets because, to my knowledge, they were not used for troop transport, troop support or attack, only medical evacuation. Combat veterans for more recent wars

...

and operations will probably also get triggered by the sounds of helicopters.

The bottom line is: helicopters have many extremely emotional memories associated with them.

Tree-lines trigger Vietnam veterans back in America because the enemy used them to hide themselves in them for ambushes in Vietnam.

I imagine similar tree lines in America might have triggered World War II veterans who fought in France around the infamous hedgerows if they saw similar situations back here in America. Unfortunately, they were so unaware of what was triggering them AND none of them talked about it.

I knew a Vietnam veteran in mid-Missouri who "went off" every springtime simply because the leaves started coming out on the trees. He got so escalated he had to be hospitalized every springtime, in spite of the multiple psychotropics he was taking. He had been a member of an "A-team" ("A" stands for assassination) but had been ambushed himself. He'd been shot four or five times, once about an inch from his heart with an AK-47 round. I have no idea why that shot didn't kill him. He also got blown up by one of his one Air Force's 500-pound bombs that people told him was a dud. It laid there for weeks, unexploded. Everyone else had been walking around it for weeks. It hadn't exploded. He walked near it. It blew up, blowing his intestines out his body cavity. I have no idea why that didn't kill him.

Green colors weren't the only triggers for him. The "closed in" experience that the leafing of the trees produced was also triggering him, reducing his distant vision.

I have worked with many combat veterans from different wars that got very upset by children of certain ages or by children crying.

One Vietnam vet I worked with told me he was getting very agitated around his four-year-old son. He did understand, after talking with him, that I didn't see anything about his son's behavior that would be reasonably irritating, especially to the level of emotion the vet admitted. We continued to process his combat experiences further until he reported that he had reflexively shot a four-year-old Vietnamese boy. The boy had walked toward him, crying as he walked. Something about the situation spooked the vet and he drew his pistol and shot the boy. The boy blew up. He had been booby trapped with explosives by the Viet Cong who were using him as a suicide bomber.

The vet then understood why he was escalating. His own son had just had his 4th birthday party. The Vietnamese boy may not have really been four years old, but the veteran had set that age in his own head and "four years old" (rightly or wrongly) was the trigger.

World War II and Vietnam War vets (and later, I bet) also get "triggered" by children crying. This is because these two wars (and the Afghan war, now) swept through civilian areas more than other recent wars America has been involved with. Many women and children got killed during the street and hamlet fighting.

I had one Korean War vet become aware that loud talking really set him off (because they were all yelling and screaming during hand-to-hand combat).

Desert Storm and Enduring Freedom Vets will be no different about the children. Unfortunately, it is also a "close-in" war with close mixture of soldiers and civilians, adults and children.

I just read about an Iraq vet that got triggered back in the states by someone simply walking toward him. When that happened in Iraq, it was often a suicide bomber or an "insurgent" assault.

Lots of people are going to walk toward you when you come back to the US, guys. Lots of them…but they are very unlikely to be suicide bombers, we hope.

…

I've also heard of other novel triggers for the OIF/OEF troops:

a) skateboard sounds-the clatter when they are dropped is close enough to clatter of an AK-47 being dropped to trigger the vet's hypervigilance. This is novel for the OIF/OEF vets because skateboards weren't developed before these conflicts. This is also novel in that the American standard assault rifle (M-16 and variations) is made more of plastic and does not make the same loose clatter sound when dropped.

b) traffic slowing down-OIF/OEF vets especially panic because when the convoy slowed down ahead of you that meant there was trouble ahead and their main defense against ambush was DRIVE REALLY FAST and over anybody in the escape route!!!!!!!!!!!!!!.

This wasn't a big an issue as a group for soldiers of previous wars because they were a) they were in the trenches in World War I, b) they were jungle wars or b) the roads in Europe and the Pacific islands were either small, muddy or clogged with slow-moving, usually friendly traffic.

There are other "treatment" issues for combat veterans that other classes of PTSD victims don't experience. The most frequent issues they face are "survivor guilt", loneliness and hatred of authority.

1) "**Survivor guilt**", I think, is self-explanatory. They feel grief and guilt about surviving when their comrades have died. They also felt (and feel) grief that they didn't "finish the job" (and feel angry at the politization of the conflict). Many of them re-enlist to "not leave any of ours on the battlefield." Many of them, now, are angry because Desert Storm didn't "finish the job" in Iraq. Many of them have re-stimulated escalated over the conflict with Iraq and Afghanistan. They also feel grief about leaving their dead and wounded comrades on the battlefield. This happens in spite of the "no man left behind" slogan of the Marines.

This grief has been going on forever but I heard one happy ending. It involved a patient of mine that was in a firefight in Korea that ended in hand-to-hand fighting. He saw a member of his patrol get shot in the back of the head by a north Korean or Chinese bullet. Blood and brains splattered everywhere. The man dropped and was motionless. My patient thought his buddy was dead. My patient was quickly knocked unconscious as a man he killed with his rifle inches away from him fell dead on top of him. He awoke some time later in another spot. His patrol member was nowhere to be seen and my patient didn't bother wondering about him because he was surely (he thought) dead.

However, months or years later, my patient saw this same man walking down the halls of a hospital back stateside. At first, my patient thought he was seeing a ghost. However, they had adequate interaction to let my patient know the other man was real and (somehow) fully functioning.

It turned out the power of the rifle bullet was simple too weak to kill his buddy.

There is another guilt some combat veterans have. That is guilt about what they did while in combat, both to the enemy and to their own. This has occurred more and more since combat has changed from the medieval charges-on-the-plains-against-massed-opposing troops to urban combat. Our current combat soldiers are still killing women and children both as "collateral damage" in urban warfare, as enemy combatants, and as irritants (intentionally killing unarmed civilians).

Killing women and children was totally "foreign" (ego-dystonic) and repulsive to Americans. I wish it were for the rest of the world, but it isn't that way in many other (if not most) of the rest of the world. I guess it isn't that way so much for Americans anymore. However, our American fantasy makes it more stunning, revolting and guilt-inducing for American troops if they do it or even see it.

...

I was recently doing an evaluation of a ten-year-old boy in a small Missouri town. After I was done, his grandfather looked at me and asked, "Is that a psychological evaluation?" I said, "yes". He then proceeded to tell me he had received one of those after coming back from Vietnam. He said he had been a Navy Seal in-country during the war and that he was aware that many of the current OIF/OEF troops were going to have it tough when they came back. He said, "because of what happened to them".

I responded, "Yes, and because of what they have done".
He responded something like,

"Yes, when you in the fight, you are so "on-the-edge" that you just kill everything that moves because it's better them than you.

He then said, "Sometimes it turns out to be a little old lady with a broom. It's too late by then. The guy who pulled that trigger has to live with it for the rest of his life, but it's too late." I suspect he was the guy who pulled the trigger but I did not ask.

I wonder how many little old ladies with brooms there have been in Baghdad, and Fallujah and Ramadi and Kabul and Kandahar and …?

I heard another OIF vet say, "You just want to kill everything". Being that "on edge" probably made for "shoot first, ask questions later" events that many of our troops will regret.

I have only heard of one "fragging"-type incidents in the 2002-2008 combat theatre (fragging-an American soldier attempts or succeeds in killing an American ranking officer with a fragmentation grenade). I have heard (very early in the war) of American Muslim soldiers attacking (and killing) other American soldier. That has not occurred since early in the war…or it isn't being reported any more that I have heard of. I could be wrong given the "fog" of war.

I have heard of several cases where American soldiers suicided after they came back from the field (from many combat theaters) out of guilt over what he had done. However, the more urban, mixed nature of more recent wars (Vietnam and Enduring Freedom), the increasing frequency of Americans killing women and children and is probably increasing combat guilt.

The strangest case of survivor guilt I ever heard of involved a draftee, non-commissioned officer who went on a POW exchange in Vietnam. His detail was picking up Air Force officers shot down over North Vietnam. One officer was panicked, paranoid and disoriented (probably from torture) and physically clung to the non-com for days. They could never persuade him to let go of the non-com. They eventually feared the non-com would die from the stress so they pried the officer's grip and forcibly separated the pilot from the non-com. The pilot died without hours of the separation. The non-com felt guilty he had let them pry the pilot off of him. He felt guilty for not holding on/out longer…as if he had abandoned the pilot and made him die.

3) **Friendly-fire guilt**: one of movie myths of war is that we never kill our own. That is probably false since the inception of mass warfare. The history channel has been informative for me by showing the cloud of arrows and spears and other hurled projectiles that occurred even in ancient battles of the Middle East. They weren't doing one-man, one shot even back then, the air was filled with projectiles to thickly that one king likened them to rainclouds.

   During the American Civil war (and other black-powder-powered engagements), the smoke of the weapons quickly caused a "fog" over the battlefield so thick that each side had trumpeters who blew out side-specific calls to inform their side where "their" side was in "the fog of war".

…

A major "improvement" reduced that "fog": smokeless gunpowder.

Unfortunately, the long-range-reach of current weapons, the closeness of support requested and attempted, and the push-the-envelope of enemy identification (night vision) continues to contribute to "friendly fire" deaths. All who know will grieve those innocent, accidental deaths, even if they don't reach the public media.

4) **Atrocity Guilt**—We are the good guys, right? We only do good things, right?

   Wrong-the "Spoils of war" have been plundered forever and it wasn't a kinder, gentler plundering. It's just getting publicized more with the increased media coverage since Vietnam when Mi Lai was published all over the media. Now the reporters are embedded.

   And with it has come increased knowledge of atrocities (from Abu Ghraib prison to rape and murder of civilian men, women and children in OIF/OEF), increased publication of it, immense guilt over it by our troops, and...suicide.

5) Another issue combat veterans face is **loneliness**. They feel very lonely in the foxhole but the continue to feel lonely and isolated from many of their combat comrades after their fighting is done because the comrades have died or drifted away. Combat creates an intense closeness (a "brotherhood") among participants (on either side). This intensity is, in part, because of the intense emotional states triggered by combat. It is also because of the intensely emotional acts of war: death, dying, mutilation, betrayal, abandonment, cowardice and heroism.

They also feel lonely because there are few civilian people they want or can talk to about their combat experiences. Who can they talk to? They (like any PTSD victim) initially want to forget all about it. Can they talk to their wives…their children…civilian peers…non-combatants? No, they don't do this because the soldier doesn't want to talk about it at all, usually from the defense mechanisms of denial, avoidance, suppression and repression. However, imagine the emotional reaction wives, children and civilians would have if a soldier told detailed events of their combat experiences. What reaction have **you** had to the combat experiences I've shared so far in this paper? Trust me, I haven't even revealed the worst I know (and this isn't even on the same emotional planet as those who experienced it).

There are many sub-cultures in our country. The military is one of those sub-cultures. Sometimes someone from that sub-culture can communicate to others in that sub-culture more effectively than those outside that sub-culture.

A warning to all in the "brotherhood": the "brotherhood" doesn't necessarily last once the war is over.

6) **Hatred/Disrespect of authority.** There are many examples of failure of leadership in the military. They are too many to detail them all. There are failures of leadership in civilian life, too, and they are, of course, too numerous to elaborate.

They all can and do produce life-long emotional reactions, be they combat or civilian failures. In the civilian world, I imagine some people ex-employees or investors of Enron will be haunted forever strong emotional surges triggered by names, sounds or other stimuli they associate with Enron, Hurricane Katrina or 9/11 World Trade Towers. They can resent the failures of leadership of Ken Lay and associates, the New Orleans levee builders and designers, or the counter-terrorism agencies who had clues but failed to intercept the suspected terrorists.

Combat veterans also have their public and private failures to lead. Those can be the generalized and erroneous (Vietnam was a "black man's war"), Robert McNamara (John F. Kennedy's secretary of defense) admissions thirty years after the end of the Vietnam War or public and incorrect (fabrication of "weapons of mass destruction" data to justify our invasion of Iraq).

Although there are civilian and military failures to lead, the combat vet feels more intensely than most because he has experienced or witnessed first-hand the tragic, violent and degrading results of those errors by those who are in charge. He was feeling very strongly when they occurred. He will continue to feel very strongly about them. The wounded feel more strongly about it because they feel more strongly about all things (due to the classical conditioning of all war-related stimuli with their body's physiological responses to the life-threatening situations and assaults on their own bodies).

The Vietnam War veterans had their disrespect of authority long before the war ended. This was due to many perceived unfairnesses, such as the Army's guarantee of a specific AIT but flunking the whole class out if they needed "grunts". The student thought their job would make them less vulnerable to combat. They found out that nobody in the military is exempt from hostile action.

The current (as of March, 2015) resentments I know of include a) repeated deployments beyond the troops tolerable limits, b) inadequate medical and psychological care, c) denial of benefits due to "preexisting" mental disabilities, d) falsification of diagnosis of PTSD to save the government money, e) combat vet suicides, and f) denial of benefits for shrapnel or other war-specific wounds (ex, from RPGs, IEDs) that the VA said are "not service connected). I wonder where they got the RPG. From Wal-Mart, Price Chopper, Kroger's?

In the mental health arena, the wounded vets are being denied compensation for supposedly pre-existing mental conditions of "Personality Disorder" (Chapter 5-13), Bipolar and even childhood Learning Disabilities.

I have several concerns about these denials.

a) I have not heard of any systematic diagnostic regime or battery of tests being given. There are "standards of practice" in psychology for adequate diagnosis. Accuracy of diagnosis without the use of standardized and normed tests is significantly below those standards of practice for licensed psychologists.

b) The current "state of the art" for most mental health practitioners requires the use the "biopsychosocial" model to understand emotional, behavioral and social function and dysfunction. That model involves taking data from the a) biological status of the subject/patient/client, b) intrapsychic processes in the here and now, and c) social environment they grew up in, including childhood and adolescence.

   This broad-spectrum, broad ranging examination covers the gamut of significant prenatal, childhood, adolescent, young adulthood and late adulthood events that significantly impact present and future level of functioning. This information is also acquired by people who know or knew the patient/soldier in his infancy and childhood. They serve as "collateral informants" to support or refute the patient's claims.

I have seen many reports in the news media that troops were diagnosed without talking to anyone in their past.

c) There are eleven personality disorders listed in the DMS-IV. None of those disorders are incapacitating in-and-of themselves. There has to be parallel behavioral evidence demonstrating a level of

...

significant social and/or interpersonal misbehavior before most of society's judicial system takes that diagnosis into account.

For example, someone might meet the criteria for Antisocial Personality Disorder. (It was previously labeled "psychopath"). While that label implies significant antisocial attitudes, you can have all of the antisocial attitudes you want (like the old "hippies") and still be positively contributing members to society. Most hippies held jobs, paid their taxes, didn't abuse their spouses and kept their children clothed and fed.

You can be as antisocial as you want as long as you don't commit a criminal act. Then the lines have been crossed and you will be asked to pay.

Another example: Borderline personality disorders are notorious for emotional surges, have immense fears of rejection, and have extreme sensitivity to criticism. You can have all those things but command others in battle or from the rear lines. You are "appropriate" as long as you don't fail your mission or perform a court-martialable act.

Another example in another diagnostic category: someone can be an official "alcoholic" (either abuser or dependent) and still honorably serve in the armed forces.

d) The symptoms of Bipolar disorder overlap the symptoms of PTSD 100% (see the DSM-5 criteria for both). (FYI: the childhood symptoms of PTSD mimic all of ADHD's symptoms, too). However, the assumption for Bipolar is that its basis is a biochemical imbalance in the brain and that there are no environmental causes. The criteria for PTSD immediately declare the symptoms are caused by exposure to a life-threatening event.

I would think a valid diagnosis of Bipolar would have to show preexisting proof of this biochemical imbalance without any environmental traumas.

e) Is there any official diagnosis of or evaluation for any of these disorders pre-enlistment or pre-placement in a war zone? Often none that I see mentioned and I have seen the military give waivers to people that have substantial problems before they enlist when the military has great need for more bodies.

f) Why does the military not have a standard psychological battery to assess for all of the issues? I have developed a battery that covers all of these issues and initially takes about two hours to administer. The follow-up interviews with collateral informants can take a few more hours at most. Compare this with the random assessments they appear to be doing now that takes months and/or years of haggling to resolve.

(PS-This lack of uniformity and standardization of assessment also applies to applicants and decisions for mental disability with the Social Security Administration).

(PPS-the VA does have standardized DBQs (Disability Benefits Questionnaires) for Mental Health, Eating Disorders and TBIs. However, these are questionnaires with no initial empirical testing). See the C&P Clinician's guides (**https://view.officeapps.live.com/op/view.aspx?src=http://www.ngwrc.org/docs/Help%20for%20your%20Claims/cliniciansguide-1.doc**) for those.

Why is the government paying billions of dollars to train and employ psychologists if they aren't using their psychological tests to their full potential? Psychological testing has a history of 4000 years. It has been refined for 4000 years. The tests can assess for everything possible, whether it is the individual therapist's favorite class of disorder or not. They can assess for things whether the therapist knows about them or not. They can assess for things whether they are the "in" diagnosis or not.

...

g) Government falsification of diagnosis to save the government money: There is a documented case of a woman psychologist in a Florida VA hospital who emailed her superiors suggesting the PTSD diagnosis be avoided to save the government money. Copies of that email went public.

In 2013, it was revealed that in 2011 a psychiatrist on the staff at Joint Base Lewis-McCord in Washington stated he had advocated avoiding the PTSD diagnos is to save money. He noted that a veteran could collect around $1.5 million in compensation and pension over their lifetime. He did not note what a healthy college graduate in this day and age could make in their lifetime. He did not note what he thought adequate compensation for loss of a limb or normal life was worth. http://news.ca.msn.com/top-stories/army-withholding-findings-of-madigan-ptsd-probe (accessed 3/11/13).

But anyway, how do you handle the combat vet's resentments and angers? How will you handle the combat vet's angers? That will depend on your own political point of view, therapeutic stance(s), level in the treatment chain, lever in the command chain, and personal issues. It will also depend on the veteran's political, philosophical and personal issues. There will be no "right" or "complete" response. I only warn you that everyone in this country needs to be ready to deal with any and all of them for the rest of the veteran's life.

Anger-constant **anger, all the time, everywhere…for the first two years.** You (the combat vet) won't know this, but it's just because you are still close to "on patrol" level of alert because of your adrenalin sensitivity and getting triggered so much (plus whatever physical pain you carry that day) puts you still on a constant adrenaline feed.

Watch out troops…your government, your family, your friends…they aren't your enemy, now. You're just overly sensitized to the stress hormones. Ride it out, grit your teeth for the next two years, pray to God, unload your gun and talk to anyone that will listen. You're safe, now, even though it doesn't feel like it.

I recommend the Iraq War Veteran's Organization web group,addressed **http://groups.msn.com/IraqWarVeterans**. This site has much information for vets and their families. I also recommend the Veterans of American website, the VA administrations PTSD webpage and PTSD Combat: Winning the War Within website.

Many people out there know what you are going through. Any Vietnam, Korean or WWII combat vet knows.

6) Moral Injury: Where was God that day? I expect all men and women faced with combat are confronted with many existential questions. I imagine all of them evaluate the meaning of life, the meaning of their life, death and the meaning of their presence "in harm's way". The answers to those questions and the means of reaching the answer(s) (or entirely avoiding the questions) are as varied as all the theories of existential questioning, spiritual searching, religious affiliating and pathological defense mechanisms combined.

However, a repeated experience of combat on combat veterans is a partial or complete loss of religious identity or faith in God.

This often is caused by events in war of the maximum horror given the American perspective of being the "good guy". The American "good guy" doesn't hurt women, much less kill them. The American "good guy" doesn't severely hurt children, much less kill them. The American "good guy" doesn't dream that a woman will try to kill him. The American "good guy" doesn't dream he will have to kill them in self-defense. The American "good guy" doesn't dream that a child will try to kill them. The American "good

...

guy" doesn't dream that he will have to kill a child in self-defense.

But in all wars since WWII, the American soldier has faced these violations of the old time "rightness" of combat. Ignore the fact that Sherman lay the civilian south in ruins during the American civil war. Ignore the fact that Julius Caesar and Alexander the Great and Genghis Kahn and …(you name them) enjoyed the "spoils of war".

The average American grunt still, I think, believes they are "the good guy", are there to help people, are going to kill or "take care of" the "bad guy" and then go home to the wife and kids and live happily ever after.

For some, that happens; for some it doesn't. And when it doesn't, they often feel they have sinned in the eyes of God or that God has let them down.

This is where the spiritual aspect of counseling or therapy is at its greatest level. You become their confessor, priest or forgiver…if they allow you to. If they don't allow you to, they will turn to fellow combat veterans, their priest, alcohol or drugs, …or suicide.

7) **"Dual Diagnosis"** Issues-the concept of "dual diagnosis" means that you have a "mental" problem AND a substance abuse problem. This, for the therapist, compounds the problems with case management since the substance abuse creates high emotional lability (mood swings), irresponsibility (The Games Alcoholics Play), and legal quagmires.

8) **Check their DD214**: The DD214 is the official discharge paper given to any soldier. It also details (since the Vietnam Era) any operations, medals and type of discharge.

Unfortunately, I have heard several men from the era of WWII up to Vietnam claim to be combat veterans, claim to have done combat, but were never actually in the military.

Obviously, you, the therapist, don't have to check to verify their military status if you are working in a Veteran's Hospital to verify their enlistment in the military. However, keep in mind, verification of enlistment isn't proof of any claimed heroics.

I almost always find that true combat veterans are very reluctant to talk about specifics. In fact, it's part of the diagnostic criteria, remember C1, above (persistent avoidance of thoughts and recollections of aspects of the trauma). It is also pretty easy to detect a pretender if you know much about weapons and ballistics, much less had real experience in combat.

So how do we, the therapist, get them, the combat vet, from "now" back to "then" when it is relevant? When do we, the therapist, get them, the combat vet, back to "then" when it is relevant? By using the following general procedure, "You did ________ (the current problem or problem behavior) because you felt ________ (the current emotion). When did you first feel this way? Is this feeling a familiar one from your past? Did this feeling occur before you were in combat? Did you feel this way before you went to war?

The patient's answer will take you wherever he/she needs to go from there. If it takes them back to childhood, so be it. If it takes them to (a) specific combat situations, so be it. If it takes them to any event related to their wartime experiences, so be it.

The Talking Therapies

2) **Individual therapy:** I have seen talking therapies provide much healing for trauma victims of all sorts. They can accomplish what no medication can provide

...

(but vice-versa, too). The following are some significant common factors to work on . There are a myriad of individual talking therapies. Some of them are generic, some of them are specific such as Eye Movement Desensitization and Reprocessing (EMDR) , Emotional Freedom Therapy (EFT), Traumatic Incident Reduction (TIR), Skill Training In Affect Regulation (STAIR), Systematic Desensitization, or Gestalt Therapy. In my opinion, all of them are Systematic Desensitization with additional bells and whistles that are probably irrelevant to the actual healing process. Refer to their books or websites for specifics about each one of them but note my previous sentence. However, note that any psychotherapy involves desensitization created by repeated exposure to the same situation or stimulus. All of that works because the individual nerves, much less the entire nervous system, decreasingly responds to the same stimulus over time. Neurophysiologists call it "habituation". Habituation serves to filter out irrelevant, unimportant stimuli. That gives the nervous system more energy and more clarity to process important stimuli.

You experience habituation every time you enter a room with a ticking wall clock. You hear it at first but you quit hearing it soon because the sound it constant and meaningless even though the sound intensity of the ticking is exactly the same as it was when you entered the room. Habituation would also occur to more complex stimuli such as your favorite bowl of ice cream. Imagine eating your first bowl of your favorite ice cream. Remember the wonderfully intense tastes and coldness of it. Now, eat 30 bowls of it. By the end of the 30th bowl, you could not tell what flavors it had, if it was hot or cold or if it was the cardboard container it came in due to habituation. Habituation (and therapy) works the same way with complex stimuli.

Discuss/Define "Triggers"- sights, sounds, smells, tastes and touches that both remind you of your combat AND, at worst, set you back in full "kill-or-be-kill" mode. Less intense emotional surging from a trigger can be extremely high but not in a full kill-or-be-kill mode. It can be seeing something that looks like your combat; **or** hearing something that reminds you of the fight; **or** smells like the fight, or tastes associated with the war, **or** touches (areas of the body, tactile sensations. Those stimuli now set off ("trigger") intense emotional surges.

The importance of "triggers" in treatment of combat vets is as important as the trigger is on a real gun.

In a real gun, you can have bullets loaded in the chamber, you can have the barrel pointing anywhere you want, and you can have the trigger cocked, but if you don't pull the trigger…it's just a useless paperweight.

You can have all the stimuli (events) happen to the vet in his post-combat life and everything is fine and dandy until… something pulls his/her trigger.

It will be most concrete (and therefore helpful) if you have the combat vets chart their triggers out. Have them describe the trigger, then describe their reactions (behaviorally and emotionally). I have included the obvious and unobvious a "triggers" that I am aware of for Vietnam Veterans and OIF/OEF veterans in the chart below. Each vet will have their own.

…

Sights	1. Seeing a helicopter (Nam & OIF/OEF): I feel tense, want to hide, I start looking around for snipers, start remembering insertions and extractions; used for "interrogation in Vietnam" .
	2.Seeing Humvees (OIF/OEF)-the all-purpose, inadequately armored "jeep" replacement make me feel …
	3. war movies…
	4. tree lines-snipers and enemy groups hid in them in Vietnam and the "Sandbox".
	5. trees budding out in the spring-made visibility much less with higher risk of ambush in Vietnam…
	6. Shopping store aisles-reduce visibility around you; hides sources of noises. Especially bad on "tunnel rats" in Vietnam. Can be very bad for OIF/OEF vets who did house-to-house searches in towns.
	7. Open spaces, like parking lots-makes one vulnerable to sniper attack because they can see you.. You feel...
	8. People walking toward you-they can be suicide bombers...
	9. My daughter/child-can remind the combat vet of a child they killed or saw killed...
	10. Intersections-frequent spots for "insurgent" fire zones in urban warfare…
	11. Middle-eastern robes: you feel…
	12. Angry people talking on cell phones…
	13. Automobile windshields-can trigger memories of bloody windshields in insurgent or innocent civilian cars after they are shot up.
	14. People driving slowly…panics OIF/OEF vets who were in supply convoys. When they were ambushed, You had to slow down to get around the damaged vehicles and wounded/killed but then s surviving vehicles drove as fast as they could to leave the area. Having to slow down over there meant there was bad stuff happing ahead, really bad stuff. 15. Overpasses-ambush sites in Iraq and Afghanistan 16. "Hanoi Jane"-Jane Fonda is an American "actress" who protested the Vietnam War by making anti-war speeches. Even worse, she went to North Viet Nam while the war was still going on and made pro-North Vietnam, pro-Viet Cong statements that were broadcast to all the world. Her statements caused some American POWs to be more horribly tortured and probably killed.

	17. Trash bags or piles of trash by the roadside-used to hide IEDs by insurgents. 18. Freshly disturbed dirt by the side of the road-also signs of a newly-planted IED. 19. Humming bird attacks-a man who was a medic in Vietnam said the red tracer rounds coming up at them when they were riding helicopters into and out of combat looked "as big as basketballs" even though they were the regular size of an AK-47 round (about half an inch in diameter). They were very anxiety evoking them because they showed you how close the enemy rounds were coming to you. He got a mini-trigger when a ruby-throated (red throated0 hummingbird buzzed him when he was on his front porch in Indiana. The *ruby* color flashing at him was close enough to the tracer color to trigger a similar excessive startle reaction. 20. Rice paddies-where are they in America? Arkansas, Washington state and other states. 21. Dead, bloated bodies floating face-down in water-seen by Vietnam and OEF/OIF vets. Seen back here by National Guard combat vets if they deal with severe floods or hurricane-related surges. 22. The red laser pointers-reminded one Vietnam vet of those red tracers. 23. Guard rails along highways-insurgents in Iraq and Afghanistan planted IEDs behind them. 24. Sponges floating in water-reminded Veterans of the bloated, dead bodies floating in the canals of Iraq and after Hurricane Katrina. 25. Roundabouts-those slow-you-down circles. What's wrong with those? They were frequent ambush spots in Iraq and Afghanistan.
Sounds	1. I hear a skateboard dropped. I feel tense, start monitoring the street around me. I wish I had a gun. Sounds too much like an AK-47 getting dropped.
	2. Fireworks
	3. Helicopters-especially Hueys
	4. Thunder, lightning
	5. Children crying
	6. People yelling
	7. Balloons popping
	8. Music associated with that war
	9. Gunfire while hunting

...

	10. Gunfire while target practicing
	11. Champagne bottle corks popping
	12. Tornado alert-sirens: also used to alert Desert Storm soldiers about possible incoming gas-filled shells
	13. Squeaky hinges on doors-everywhere in Iraq and Afghanistan.
	14. Asians whispering-done by Viet Cong, NVA and Japanese soldiers to "mess with" American troops. May or may not have preceded their attacks. 15. Too quiet-children and dogs would vanish from the city streets when they knew there was an ambush set up. 16. Electric wheel-chair motors on startup: a paralyzed therapist told me his electric wheel-chair motor reminded Vietnam veterans of the electric motors driving their gun swivels. This sound would/could also trigger any modern-day, electrified combat veteran who had been around motorized guns. . 18. "Hanoi Jane"-see visual "trigger" section, above, for discussion of this "actress" and how she got this nickname.
	19. Alarm bells-hospital alarms triggered an old WWII Navy combat veteran to scream "She's going down, she's going down" (like his ship did when it was torpedoed).

...	20. Popcorn popping-sounds like automatic weapons firing. 21. Crowds-too overwhelming, too much noise coming from everywhere. An Iraqi vet, amputee, got a big planned welcome home party by his hometown in Indiana. He had been in rehabilitation for many months due to his severe physical wounds. The town sent an escort to bring him home on Friday with the big parade scheduled for Saturday. He got emotionally overwhelmed by the initial Friday escort crowd, got intoxicated, fired a weapon (didn't hit anyone), got arrested and they had to cancel the welcome-home parade on Saturday. 22. A fishing rod whooshing sound as it whipped out the fishing line-sent an Iraqi combat vet under a table because it sounded too much like the incoming sound of artillery. 23. "Catch the baby"-this fun game is often played by proud relatives of a new baby much to the delight of the adults and child. However, a medic from the Vietnam War told me the Viet Cong would wrap a bomb or booby trap in a baby blanket and toss it to an American soldier saying, "Catch the baby" and the American soldier always would catch it because he thought it really was a baby. The booby trap went off, killing and/or maiming him. 24. In a similar vein, during OEF/OIF, females in Iraq and Afghanistan would walk up, hand a real life baby to an American soldier and hurry off. The Americans sooner or too late found out the baby was wired with explosives and could explode. On a unexpected note, one soldier told me he unexpectedly astounded that the wires were wrapped around the baby so tightly that it cut into its flesh. This triggered Moral Injury. 25. The song that was playing on the radio of your military vehicle when it got blown up. 26. "Thud"-that sound when something heavy and soft hits the ground. Triggered a Vietnam vet back home. Reminded him of the sound the VC's bodies made as they hit the ground after being thrown from a hoovering Huey during "interrogation". I've heard of many who saw the bodies go out but never heard of a soldier who heard them hit. To make it worse, he reported the ROK (Republic of Korea) soldiers took four VC up for interrogation. They threw the first three out pretty quickly. The vet heard them scream as they fell, then heard them hit the ground, killing them. The vet said the fourth one was in the helicopter for quite a while before they threw him out. That meant he gave them some valuable information…before they killed him. 27. People whistling-sounds too much rockets or mortars as they fly in or fly by.

Sounds (continued)	28. Ice cubes tinkling in a glass-he hasn't figured that out yet because he served in Iraq where there was no ice tinkling in a glass where he was.
	29. Billiard balls breaking-he never figured out why that triggered him so badly but he sold his billiard table.
Tastes	1. Foods of that era…but typically not eaten since the smell is experienced and produces revulsion before they get the chance to eat it.
	2. Many WWII American veterans of the Pacific theater never ate Chinese food after the war.
	3. Many WWII American veterans of the European theater never ate German food after the war.
	4
	5
	6
Smells	1. Burning plastic 2. Foods 3. Exploded gunpowder 4. Incense 5. Blood 6. Fuel oil 7. Diesel oil 8. Raw sewage 9. Meat burning on the BBQ-smells too much like burning human flesh. 10. Decaying animal odor-smells very much like decaying humans.
Touches	1. I particularly want to note that I have talked to two combat vets who were grabbed from behind and responded with lethal combat maneuvers…and killed some stupid drunk in the bar who tried to be "tough" with them. One spent many years in prisons for second-degree murder. I'm not sure the other one was being truthful about it or was stealing valor. 2. Lawn mower vibrations-I had a Vietnam Vet Huey helicopter door gunner tell me he finally (years after coming back home) became homicidal every time he mowed the yard with his push lawn-mower because the vibrations of the machine set off his "kill-or-be-kill" feelings as a door gunner.

	3. Motorcycle vibrations-same possibility of "triggering" kill-or-be-kill as a lawn mower. Could be the reason for the ferocity of the Hell's Angels. Their first chapter was formed by a group of ex-WWII Bomber squadron survivors. They liked the sound of the Harley-it reminded them of the sound of the airplane motors. They may have also been taken back to "kick ass" attitude of a bombing mission by the vibrations of the motorcycle engines. The steering wheels and .50 caliber defensive guns in the bombers vibrated a lot. The gunners in Humvees and all other weapons with heavy vibrations (SAWs, .30 cals, M2s, etc.) may also be susceptible for "triggering" by lawn mowers, motorcycles, jackhammers, etc. 4. Hands in cold water-an American WWII sailor floated in freezing water for three days after his ship was sunk in combat. He heard and saw the sharks attack and kill many of his shipmates in the water with him. He waited for his turn to come to be attacked for those three days but it never came. The rest of his body that was totally submersed went totally numb quickly. His hands, being only partially submersed and clinging to things to help him stay afloat, kept feeling the coldness of the water. He thought less and less about those three days over the passing decades until his wife brought home some fresh shrimp and wanted him to prepare them for cooking. He had them in icy water to peel them and flashed back to the feelings and memories of those three days in the icy, shark-infested, deadly waters. 5. Coldness, itself-tough for Korean War Veterans to deal with since Korea was very, very cold at times. Also difficult for Vietnam veterans to tolerate if they were in the monsoon rains for any length of time, because hypothermia can set in even at temperatures of 55 degrees. 6. The heat of summer stateside-is difficult for veterans of OEF/OIF who fought and died in Iraq and Afghanistan since the average temperature in those countries is 130 degrees F. 7. Stepping on squishy things-a Navy man I dealt with was traumatized by a death-defying onboard ship mutiny so he was primed for "fight or flight". He never had to react any more during the war or his military service but when back home a young girl was severely injured by a toothpaste-bomb booby-trap set in a major box-store customer-service

...

	counter. It blew her hand off, scattering parts of it all over the floor. He, being the warrior, rushed to help with the problems and pick up body pieces. He stepped on parts of her hand as he was cleaning it up. He would severely trigger years later when he stepped on other squishy things like grapes on the floor or banana peelings.
	Proprioceptive: sense of body position relative to the rest of the body (what they test with the drunk driver "close your eyes and touch your nose" test): one Vietnam vet I heard about triggered after being handcuffed and his hands raised above his head. This body position imitated the body position he was in as he was brutally tortured in a bamboo cage in the jungle of Vietnam (burning sharpened bamboo stick shoved up his anus). A second Vietnam combat vet was captured, slung on poles with hands tied together, feet tied together, carried for many days, beat repeatedly, rescued just accidentally when another American squad passed by. After he came back to the United States, many years passed and he went to a carnival with his grandchildren. One of his granddaughters strapped him in for an exciting ride and he got triggered and the look on his face frightened his granddaughter because she said "You looked like you wanted to kill me."
Time	(Anniversary Reactions) 1) The same month of the year that a trauma occurred. The person starts getting more uptight about one month before the trauma happened and gets increasingly uptight until they either explode, figure out the connection or the date passes. 2) The same season of the year that the trauma happened (see the tale of the Vietnam vet who was my farrier. He would get really tense every springtime when the foliage came out and he could not see as far out into the distance. He got so tense he would have to be hospitalized on a psychiatric ward. 3) One WWII veteran of the Battle of the Bulge told me the Germans would shell the Americans every day at 10 am and 2 pm. He said, "and they did it intentionally!" Of course they did.

I had a patient who was wounded in hand-to-hand combat in the Korean War. I was working with him at the time the movie "Saving Private Ryan" came out. One week, at the end of our therapy session, he enthusiastically told me he was going to see it. I vaguely discouraged him but also told him to note how he was feeling during the movie. (I knew the sounds of battle in the movie were going to be "triggers" for his own PTSD).

During our session the next week, he mentioned he had seen the movie. I asked him how he had been as he watched it. His initial response was comments about the movie. I stopped him. I said, "I didn't ask you about your impression of the movie; I asked you about how it affected you". He admitted he had caught himself huddled up in the chair in a fetal-ball as he could get when battles raged on the screen. Those noises in the movies were like noises paired to him in reality with horror, survivor guilt and immense pain

I had a friend who was a combat veteran of Vietnam. He was a dog handler and often led "point" on patrol. He and I saw a Star Wars-type movie together one time several years after he came back home. There were enormous explosions on the sound track, emphasized by as-big-as-they-could-get speakers. I knew he was a combat vet. However, he came out of the movie unshaken. He actually enjoyed the explosions. I couldn't figure out why he wasn't all stirred up by the similar sounds. Years later, I figured it out: He had never been wounded. He had been through enormous firefights. He said he was in one firefight where 10,000 rounds (bombs, artillery, rifles, mines) were fired by all sides in five minutes (his approximations). He had seen many killed and wounded. He, himself, wasn't ever wounded.

The sounds, although similar to the bombs, artillery and mortars that had dropped all around him during firefights, were not paired with the immense pain of wounds for him. They remained…just loud sounds.

…

You, as a therapist, need to set up a chart to help your combat clients have concrete examples and displays of their "triggers" with old and new coping responses. That chart could look something like this:

"Trigger"	Old Response	New Response, Daytime	New Response, Nighttime
1. Heard a helicopter	Feel tense, have combat memories, go inside, start to drink	Go inside, call my friend, meet them in the park.	Go inside, call my friend, go work out with them.
2			
…n			

This charting may appear to you to be too complex or overdone. However, the veterans need to start being able to do this in their head as soon as possible and they can see it better internally if they can see it more externally.

In addition, I recommend you push this "trigger" charting to the limit. Keep an ongoing (rolling) chart. Each client/patient will see more/different things as they progress through the forever-after journey.

Atrocities to women and children: in addition to the many triggers above, I have been told of three situations that American soldiers witnessed that traumatized them. They were:

1) A Vietnam veteran "tunnel rat" came upon the body of a young Vietnamese girl whose breasts and vagina had been cut off/out by the Viet Cong because she had (or they suspected she had) fraternized with the enemy.
2) A group of American soldiers came upon a group of pregnant Vietnamese females who had been hanged and their dead fetuses (American bred) had been cut out of them and were dangling from the mother's bodies by the umbilical cords.

3) American medical personnel have always tried to help the native people. However, soldiers from Iraq and Afghanistan told me that the insurgents killed wounded children that American doctors had saved the day before and left them outside the medical facility the next day in an spot the medical personnel were sure to find. This was done as a warning for the natives not to accept help of any kind from the enemy.
4) Religious-based fanatics are raping and killing men and women's spouses in front of them.
5) Religious-based fanatics are raping and killing children in front of their parents.

Thusly is the barbarity of war.

COMBAT VETERAN

TRIGGER TRAUMA THERAPTY (CVT3)

I would like to emphasize to the reader a novel treatment approach I seem to have developed that I call Combat Veteran Trigger Trauma Therapy (CVT3). I feel it is superior to all other combat trauma treatments because of its applicability and clinical effectiveness for the combat veteran as they deal with the real world once they come back to the supposed safety of the USA. It involved the traditional "working through" of a trauma by the patient sharing it with the therapist. However, it then focuses on giving the combat vet a useful analysis tool of the lingering effects that trauma from "over there" can still have on them "back here". It has five different aspects. They are:

- 1) Have the vet narrate or write down a traumatic event.
- 2) Work with them to identify all "triggers"
- 3) Have them do homework to identify the triggers setting them off in their day-to-day lives.
- 4) Have them practice effective coping skills when triggered, including decompression (relaxation,

...

calming down, reframing, doing manual work to wear out their adrenaline/cortisol driven jitters), even after reviewing the triggering events with you. Practicing these decompression activities prn (as needed) instead of inappropriately displacing that tension on innocents.

- 5) Share known triggers with the vets support system, both family and institutional (24/7/365)

I would like to give the reader an example of how to identify the specific "triggers" of a combat vet's recounting of his combat traumas by watching and analyzing a film clip about survival guilt. I refer the reader to the filmed account by Staff Sgt Adam Lingo at www.spike.com/collection/15805?cmpnid=800&lkdes=COL_15805.

My respects to SSgt. Lingo, his men, their sacrifices and their efforts. A transcription of his narrative is included below for analysis purposes and in case the reader has no access to this web site or the internet. The producer of the clip identified SSgt. Lingo's problem as "survivor guilt". I analyze the same clip for "triggers". I have never asked Ssgt. Lingo for his permission to use this clip because 1) I don't really need to do so since it is public domain material and 2) I don't know how to contact him. If you do know him, let him know I ask his permission to do so out of respect for him.

He said, "It was April 5th. I'd received instructions the night before and I was gonna take my squad and one of the sniper teams. We were gonna set up an overwatch by the road that was constantly getting hit by car bombs and IEDs because we had specific "intel" that there were gonna be attacks that day and it was a road constantly used by our convoys…so we were basically there for security so if anything suspicious or actually we caught guys trying to put IEDs in the curbs or whatever, we could stop them.

So anyway, the whole day started out from the minute we got there, it started to go bad. It was the one time I went on a mission like this that I carried two radios and it was a good thing because I started having problems, "commo" problems, radios didn't work for me...and so-on and so-forth. But, we inserted during the night into a building so no one would know we were there and our mission was to watch the ad and look for anything suspicious. We managed to stay on the roof for the whole day without anyone knowing we were there which is pretty hard to do in a built-up city. We had been watching around (the city) most of the day.

Right around noon, I know it was noon because I could hear the mosques call for prayers going off, we got a call on the radio that there was specific "intel" that there was going to be an attack. They weren't specific on what sort of attack but the one thing we had learned, after getting hit several times, was to watch out for car bombs, most of the times drive by suicide bombers. So around noon, I'd just switched with one of my guys, rotating, his turn to watch thru the hole and my turn to monitor the radio traffic. Not even two seconds after I had switched out with him and sat down to grab the radio, there was what I can only describe as the loudest, most extreme explosion I've ever heard in my life. It was a car bomb. It's impossible to describe the feeling when a car bomb goes off. It's like my lungs were going to come out of my chest. That concussion. Boom! I know right off the bat when I heard that. A couple of seconds later, debris started raining down, chinks of metal flying everywhere. I knew right off the bat what had happened. One of our convoys was driving down the road and a pickup truck, it looked like any other pick up, was parked on the side of the road and as soon as our convoy came out, the pickup rammed into a Stryker and blew up.

The Stryker weights 23 tons and this car bomb was so powerful it picked up the entire Stryker and moved it, threw it into the median of the road. When I ran to the edge of the roof and looked over, I could see the Stryker, still burning. We took some small arms fire right after that. I immediately got on the radio, called the platoon out there in the convoy to see if they needed any help with casualties. Their platoon sergeant told me "no, we're better off if you stay up there and

...

provide overwatch for us just in case there is a follow-up attack". So me and my squad and snipers pulled local security just in case anyone tried to sneak up the alley ways as they commonly did.

A friend of mine, Joseph ______, was the vehicle commander for that Stryker and when the car bomb hit the Stryker, there was a small piece of shrapnel from the car bomb that went straight into his forehead and came out the back of his head. It had gone through his brain, obviously. He wasn't very coherent from what I understand but he was moving around, trying to grab his head. So, of course, they grabbed him and the squad leader for that vehicle and the driver, the three guys inured pretty badly and took them to the hospital as quickly as they could. Things didn't look too good for him. They wound up doing a craniotomy on him because his brain had been swelling and they had to remove the top of his skull so that his brain would quit swelling.

Needless to say, he'll never be the same. He survived but lost a lot of basic brain function. This happened over one year ago and he's still in the VA hospital and I feel guilty, partially responsible. Maybe if I'd paid better attention that day or something that I'd been able to stop that attack."

I think the typical therapist would feel compassion, discuss survivor guilt and leave it at that. However, I now I want any therapist, vet or family of a vet reading this book to go through the narrative, above, and underline or highlight any parts of his narrative that might serve to be a "trigger" (in any modality) for later combat vet trauma triggers. Go ahead and underline the triggers you can think of.

Now I'm going to repeat the narrative below with my own underlines to show what triggers I can see (both obvious and more subtle). Then I will discuss them all.

He said, "It was April 5th. I'd received instructions the night before and I was gonna take my squad and one of the sniper teams. We were gonna set up an overwatch by the road that was constantly getting hit by car bombs and IEDs because we had specific "intel" that there were gonna be attacks that day and it was a road constantly used by our convoys…so we were basically there for security so if anything suspicious or actually we caught guys trying to put IEDs in the curbs or whatever, we could stop them.

So anyway, the whole day started out from the minute we got there, it started to go bad. It was the one time I went on a mission like this that I carried two radios and it was a good thing because I started having problems, "commo" problems, radios didn't work for me.and so-on and so-forth. But, we inserted during the night into a building so no one would know we were there and our mission was to watch the road and look for anything suspicious. We managed to stay on the roof for the whole day without anyone knowing we were there which is pretty hard to do in a built-up city. We had been watching around (the city) most of the day.

Right around noon, I know it was noon because I could hear the mosques call for prayers going off, we got a call on the radio that there was specific "intel" that there was going to be an attack. They weren't specific on what sort of attack but the one thing we had learned, after getting hit several times, was to watch out for car bombs, most of the times drive by suicide bombers. So around noon, I'd just switched with one of my guys, rotating, his turn to watch thru the hole and my turn to monitor the radio traffic. Not even two seconds after I had switched out with him and sat down two grab the radio, there was what I can only describe as the loudest, most extreme explosion I've ever heard in my life. It was a car bomb. It's impossible to describe the feeling when a car bomb goes off. It's like my lungs were going to come out of my chest. That concussion. Boom! I know right off the bat when I heard that. A couple of seconds later, debris started raining down, chinks of metal flying everywhere. I knew right off the bat what had happened. One of our convoys was driving down the road and a pickup truck, it looked like any other pick up, was parked on the side of the road and as soon

…

as our convoy came out, the pickup rammed into a Stryker and blew up.

The Stryker weighs 23 tons and this car bomb was so powerful it picked up the entire Stryker and moved it, threw it into the median of the road. When I ran to the edge of the roof and looked over, I could see the Stryker, still burning. We took some small arms fire right after that. I immediately got on the radio, called the platoon out there in the convoy to see if they needed any help with casualties. Their platoon sergeant told me "no, we're better off if you stay up there and provide overwatch for us just in case there is a follow-up attack. So me and my squad and snipers pulled local security just in case anyone tried to sneak up the alley ways as they commonly did.

A friend of mine, Joseph, was the vehicle commander for that Stryker and when the car bomb hit the Stryker, there was a small piece of shrapnel from the car bomb that went straight into his forehead and came out the back of his head. It had gone through his brain, obviously. He wasn't very coherent from what I understand but he was moving around, trying to grab his head. So, of course, they grabbed him and the squad leader for that vehicle and the driver, the three guys inured pretty badly and took them to the hospital as quickly as they could. Things didn't look too good for him. They wound up doing a craniotomy on him because his brain had been swelling and they had to remove the top of his skull so that his brain would quit swelling.

Needless to say, he'll never be the same. He survived but lost a lot of basic brain function. This happened over one year ago and he's still in the VA hospital and I feel guilty, partially responsible. Maybe if I'd paid better attention that day or something that I'd been able to stop that attack."

In this clip, 7 minutes and 19 seconds long, I counted 15 specific triggers (obvious and extremely subtle) which I underlined above. I will now discuss each one of these triggers:

April 5th- an “anniversary” trigger. SSgt Lingo may start becoming agitated and uptight, with little knowledge of why, around the first of April, if he doesn’t realize it’s nearing the calendar date for his combat trauma.

Sniper Teams-he may become uptight and agitated when he sees television programs about snipers because there was a sniper team with him that day.

Two radios-or two of anything. If someone tries to get him to “take an extra”, \“take two” or something like that, he may become increasingly agitated.

On the roof-he may become agitated, tense, uptight if he sees a city from an elevated perspective, like if he gets up on his roof to clean the gutters, patch a roof, adjust an antenna, or even have a high perspective from a high-rise building.

Noon-he may become agitated, tense and/or uptight around noon.

Mosque call for prayers-he may become agitated, tense and/or uptight if he hears a Muslim “call for prayers”. Many big cities in America have Muslim populations and mosques. Most small American cities don’t, but that doesn’t mean SSGt. Lingo won’t visit a big city one day and hear the call to prayers and feel…

Switch off from a shift-different part, same thing regarding his feelings. He may have trouble giving up his “watch” on any activity back home and/or have intense feelings of guilt.

Watch through a hole-he may get triggered if he ever has a restricted, from-the-hole view back in the states.

Loud explosion-too obvious. Any loud explosion back home will probably make him very upset for a long time to come.

Debris raining down-it may just be papers falling on his head that someone dropped off the balcony at the library. Maybe it's a tissue someone accidentally drops off the balcony at the opera or a pencil that falls on his head at a concert. Maybe it will be leaves falling off a tree in the fall. Anything like the debris raining down can set him off, again.

Stryker-seeing one, seeing a program about one, etc., can trigger him, again.

Joseph________, his wounded friend-any pictures, voices, etc., that look like or sound like Joseph can "trigger" intense feelings for SSgt. Lingo.

Grab his head-maybe SSgt. Lingo's son, wife or any other civilian had a bad (but not life-changing head injury) and grabs their head and howls. This can trigger SSgt. Lingo into intense emotions.

"Craniotomy"-this word, uttered on a medical program, real or faked, can set Lingo off.

"Remove the top of his skull"-did you ever see Anthony Hopkin's second Hannibal Lecter's second move, Hannibal? Remember the scene in it where Hannibal lifts off the top of the man's skull and scoops out his brain? SSgt. Lingo may have intense emotional surges if he sees that scene.

I hope you, the reader can appreciate both the obvious and subtle "triggers" that might emotionally effect, haunt and drive SSgt. Lingo from this one brief episode. I hope they do not haunt him. I hope he's had received good help over this event and any other aspects of his tours in the "sandbox" if he is still bothered by them.

The therapist can should also discuss more general "triggers" Ssgt. Lingo might have from this event that he didn't mention in that clip, such as fireworks, people yelling, etc., that are mentioned in the tables of "triggers", above.

By doing the CVT3 analysis, you give any vet many specifics to be aware of during the week outside the therapy environment to help defuse their "triggers".

The reader can view other real scenes of combat, if you have the stomach for it, to identify other "triggers" of PTSD when the vet has returned stateside. Even the Iraqi sniper clips they posted on the web as they filmed killing American troops can be used. They give me much sadness to watch but my sadness is nothing compared to the emotions the men and women around those troops felt and still feel as they saw their comrades cut down so quickly and violently. You all have my respects.

An "**Anniversary Reaction**" is a type of "trigger" that is much more difficult to identify and deal with for most victims IF you don't tell them about them or they don't learn about from another source.

Anniversary Reactions are emotional and cognitive reactivation of trauma responses caused simply by the same time of day, month or year occurring. Grief therapists first identified them but they occur with trauma victims, too. I have seen most Anniversary Reaction emotional triggers set off by the time of year (the seasons or certain holidays). Holidays will trigger children more than time of the year simply because children don't have an "adult" sense of time.

I have also seen many people with great sadness from any cause(s) be more irritated, agitated, withdrawn and/or grumpy during family-focused holidays.

I believe this is due to what I call the **"You Glad, Me Sad" Contrast Effect**.

Family centered holidays are "supposed to be" times when scattered family members get together, celebrate the good fortune of the time, catch up on the good fortunes that have

...

happened since the last get-together, and memorialize the strength of the family.

Any recent or severe trauma-victim, regardless of the cause, gets their loss magnified (by simple contrast) by the increased cheeriness, happiness, celebrated prosperity of those around them.

The other aspect of this contract effect is that the Veteran might have been deployed during the same holiday, suffered mildly (homesick, minimal food) or was under attack or doing a combat mission and is thinking about what they were going through that time while everybody else back home (then and now) was just celebrating.

I find the best way to help the Contrasted griever is to acknowledge that they don't have as much to celebrate at the reveler. Don't expect much warm-n-fuzzy in return, but it will help them if you acknowledge their continued pain (without being extensively maudlin about it).

Dual Diagnoses: involves diagnoses on Axis 1 of the DSM Multiaxial Nomenclature System with a drug, alcohol or Axis 2 Personality Disorder. See specific treatment paradigms for each. I don't' want to go into detail about any of these at this point.

Drug/alcohol abuse: discuss how trauma victims often abuse alcohol/drugs to:

a. numb themselves psychologically
b. be popular and/or accepted by others (often with other abused or neglected peers)
c. knock themselves out to sleep.

Dealing with Feelings: most significantly, teach them effective ways of <u>coping with</u> their feelings, no matter what they are. Unfortunately, I have come to see a phenomenon unique to severely traumatized people that I call "***Emotional Tsunamis***". Psychiatry called them "triggered manic episodes" but that doesn't capture the "triggered" nature of the "transference" emotion.

A tsunami is a massive tidal wave triggered by an underwater earthquake. The energy of the plates slipping is transferred at the speed of sound and can flow for hundreds of miles, such as occurred in December of 2004 in Indonesia and surrounding islands.

The ***Emotional Tsunami*** I am talking about with severe trauma vets is the massive emotional surge(s) that happen after they get triggered by something now that can run full-bore for weeks, even months afterwards. They will be on combat sleep pattern, do perimeter checks, set up booby traps, sleep with a weapon, and be hypervigilant again (with some increasing drug or alcohol intake probable).

Nobody ever told me about these prolonged surges during my graduate school or post-graduate workshops. Not the vets; not the mental health professionals, not my pastor, not even my combat patients or their families.

They are real, they are frequent in combat vets for many years down the road and they require **intensive and extensive** case/crisis management. In some cases, it will require…

Suicide intervention and prevention (we hope!)-Effective suicide prevention will take "the village" (all facets of society) on call 24/7/365 "till-death-do-us-part" to adequately protect the combat vets from suiciding. Eight to five, Monday thru Friday won't work. Agony doesn't know a time clock or calendar. In fact, it prefers to sneak up when your guard is down, when they are "partying" (a euphemism for getting drunk or doing illegal drugs).

The number of combat veterans from WWII, Korea and Vietnam who suicided during and after they returned will never be known because no official records were ever kept. The number of Vietnam vets who suicided was estimated by the DOD as 38,000, the urban legend was "twice as many who were killed in combat" and some wild hares estimated over two million.

Although the current (OIF/OEF) conflict numbers was in that "fog of war" for many years, a 2009 joint DOD/VA National Mental Health Summit held in Washington, D.C., declared that "more veterans have committed suicide since 2001 than the number of service members kill in Iraq and Afghanistan over that period". Shades of Vietnam.

The current (2017 and 2023) body count for military veteran suicides is 22 per day. One a day is still in-service, the rest are out. The ones out are heavily composed of those over 50 (i.e., Vietnam vets) and female vets who were sexually abused, discharged under bogus problems (Personality Disorders, not meeting performance standards), denied any benefits so they remained traumatized, neglected, homeless and without benefits. Those do not include suicides in Texas or California because they are not legally ordered to report those (and they are two of the most populous states in America with two of the highest groups of military active and retired).

The actual branch of the service with the most suicides has changed every survey taken. First it was National Guardsmen and people thought, "Oh, they weren't professional soldiers". However, these men and women were sent into the heaviest fighting in both Afghanistan and Iraq with incredibly substandard equipment. For example, the Humvees were not up-armored at first and had only canvass sides, then plywood and sheet metal sides, both of which were incapable of blocking any bullets, much less RPGs and IEDs. These men and women were also ill-equipped (i.e., non-equipped) with any protection against IEDs or bullets because they had no body armor. Americans back home started mailing their vets plates of metal and any body armor they could find to protect their loved ones. However, the very next year the same survey found the Marine Corps having the most casualties. Nobody knew what to say. They certainly could not discredit the training or the professionalism of those men and women. Their equipment was still very substandard. The next survey found the Special Forces the highest; the survey after that it was those in boot camp waiting to deploy. One solider told me there was a wave of suicides of commissioned officers while everybody was waiting to spread out from Saudi Arabia and Kuwait. Within the last couple of years, a VA survey showed it was veterans who had just been discharged from VA mental programs. Then it was female soldiers. There have been several veterans who suicided in VA parking lots when they were turned away from inpatient VA mental health programs.

Sleep problems-helping them while keeping yourself from getting killed (we hope!)

There are a myriad of sleep problems the combat vet faces while in the battle zone and after he or she gets back stateside. These are classic and are documented across all combat actions I am aware of.

The first of these is "sleep onset" disorder-getting to sleep in the first place. This occurs with civilians and soldiers alike. There is nothing exotic or significantly different about it in either arena. You are simply so mentally preoccupied with something in your daytime reality (past or present) that you can't get to sleep. Unfortunately, the combat veteran may be re-stimulated in the "fight or flight" level of adrenalin arousal that keeps him going for days, if not weeks, on end.

The most frequent and healthiest cure for that problem with the civilian is simple passage of time. You solve the pressing issue or you don't. Your sleep will return to normal if you solve it; your sleep will stay distorted if you don't...but your body will physiologically compensate for the sleep loss by many different mechanisms. Your brain can do "REM rebound" (making up for the lost sleep at a later time when the stress dissipates) or even microsleep episodes (sleeping with your eyes open and not knowing you have gone to sleep.

Although there was a popular fascination and hysteria about sleep deprivation in the 1950s, there has to be severe and prolonged sleep deprivation for intense confusion and even hallucinations to begin.

The average day-to-day stress-related sleep problems resolve with use of effective coping skills, sheer passage of time or both. You may notice that you often "finally" get to sleep about two or three in the morning. This is because your body finally reaches its lowest level of body (circadian) rhythm and physically goes to sleep, no matter what's on your mind.

There is one simple (but effective) way to stop the train of thoughts that are often associated with stress-induced sleep onset problems. Just write the concerns down you have on a sheet of paper, write down possible solutions, then prioritize all of it. This simple procedure literally helps you get things "off your mind" and you can go to sleep more easily.

Don't take sleeping pills. I have never seen any of them really help and they are all physiologically addictive with some strange (not really) side-effects like "sleep driving", homicide while sleepwalking and homicidal rages with amnesia when mixed with alcohol.

The second simple cure is microsleep episodes which also occur with extreme and severe sleep deprivation. The combat vet's level of arousal is probably so high at this point that they can only come down with an IM (intramuscular) dose of soporific (sleep medication) or Thorazine.

The second sleep problems are the bare-faced nightmares. They are particularly vivid and riveting for any severe trauma victim. They often, at first, are pure nocturnal reliving (flashbacks) of the traumas. They can persist for decades and can be purely mental in process or (usually) include physical, mental and behavioral reenactments. The dreamer can attack or hide from the enemy all over again during their nightmare.

The level of emotion in the nightmares is also tsunami-like and can carry over into the dreamer's daytime for hours if not days.

The main thing I want to emphasize for spouses, partners, children and anyone else around the person having a nighttime flashback:

1) Don't touch them!!!!!!!!!!!
2) Don't deal with them from behind if at all possible.
3) Repeatedly call out their civilian name, reassure them they are safely back in the "world" (or use whatever slang term his combatants called being in the United States), tell them what state, town and street they are at (to physically anchor them back to this world) and that they are just having a nightmare of the war. Do all you can to reorient them to their safe here-and-now. Repeat all of this over and over and over. Eventually it will sink in to the nightmare itself. It may take them hours (or days) to decompress from the emotions of the nightmare and/or flashback. Have them talk about it, write about it, work it off, play it off.

This is prime time to a) try to keep them away from alcohol and b) keep them talking to you or someone else they can relate to about the event(s) behind the nightmare. If you can't stay with them until they are decompressed, call in support.

This is where I recommend families of combat veterans remain linked together as support system for all other issues but, especially, for periods of decompression from these Tsunamis. These periods are extremely stressful and taxing for all involved. These are the highest probably periods of danger to self or others.

Theory 1: A person who has suffered immense trauma may never completely "get over it".

Theory 2: A severely traumatized person (civilian or combat) has 58 gallons of tears to cry. The sooner they cry them out, the better they'll feel.

Combat Vets: I wrote "58 gallons" but that, of course, is an arbitrary number. My real point is: You will need to cry and

...

share and release all of your stored-up feelings and thoughts. As you do, your emotional tension subside, and you will be able to deal more with life back in "The World".

The last sleep problem involves sleepwalking. It is more common in people who are stressed. It occurs in Stage 4 of sleep. It isn't exactly "acting out" of a dream because they aren't officially "dreaming".

Sleepwalking (and sleep talking) occur in Stage 4 sleep. The sleepwalker is having some mental activity in their Stage 4 sleep. I have heard of many combat veterans sleepwalking but haven't had any type of access to them (direct or through the media) to find out what their sleep mentation was.

I recommend problem sleepwalkers set up motion-triggered lights inside their house (with additional auditory alarms included in the system) to make them alert to their sleep walking. These systems can easily be rigged from the motion-detection yard lights available at many hardware stores.

Marital Therapy: There are going to be the typical re-integration issues where stateside-based parent is going to have to learn how to share power and decision making with the returned veteran. There may, unfortunately, be some dealing with infidelity. There will be many issues of teaching the noncombat spouse about PTSD and how to cope with it. The majority of those will be:

- Domestic violence
- Child abuse
- Substance abuse
- Infidelity
- The vet's emotional distance
- Getting a job

Unfortunately, there have been some incidents of total lack of self-control and insanity that spouses have to cope with. I can give you a few examples I have encountered in my 43 years of clinical work.

For example, one 49-year-old woman I evaluated in Missouri reported that her father was in a "clean-up crew", (cleaning up our dead soldiers and body parts after battles) during the Korean War. He was a steady worker and a pillar of support for her when she had brain cancer. Her mother, his wife, said he was "changed" from the war experiences. He was very suicidal and would drive around with all of the family in the car with his eyes closed, praying to God to take them all. He also would chase her mother around the house when she was nude, beat her every weekend, broke her ribs and tore her hair out.

Did this happen to every vet and his wife after they came back? No, of course not, but I want to prepare the health care providers for the worst of the worst so that you give each vet proper 'framing' (of Duty to Warn) regarding domestic abuse.

A milder, but typical to this very day is the WWII Navy veteran who worked driving delivery trucks, did "payday" drinking binges (that lasted for 1-2 days at a time), had some DWIs and lost jobs because of his alcoholism and served in the military. He was in the Navy during WWII and talked about his ears bleeding after his ship's guns fired. This man's son reported his mother dealt with his father's binges by going to AA, talking to her friends for support and "not talking to him" (dad).

This man denied his father was "hypersexual" but admitted he thought
his father was unfaithful to his mother "maybe once".

Not uncommon, not exotic and not only done by combat vets.

Family Therapy-the most distraught of the vets will be the most abusive to their children and spouse. Some examples:

I had a 39-year-old patient whose father served in WWII in Europe. He came back "changed" and bizarre. He claimed he had been a spy (when he was just a "grunt"), kept snapping off "Heil, Hitler" shouts and salutes as he walked around town (not a

...

popular move after World War II), would spit in people's beer at the bars so they wouldn't drink it (and he could get it free), became a horrible alcoholic, and would hit the children in the head with thrown objects (from <u>many</u> yards away) if they became too noisy.

All my patient had been told, at first, was that his father had been grazed by a bullet but not seriously wounded. However, once I pursued his father's bizarre behaviors, his father's cousin told us about how his father had actually been wounded in a large explosion that blew one eye out of its socket. He stood back up, pushed it back in his eye socket and kept on fighting.

Given that information, his bizarre behaviors sound more like Traumatic Brain Injury (TBI) in a soldier that many did not know could survive such (invisible) wounds.

I worked with a 58-year-old Vietnam War combat vets whose father was a WWII combat vet. His father was a "scout' in the Pacific theatre and got shot in the back of the head with the bullet exiting one eye. It didn't kill him but produced what we now know as TBI. How did he behave when he came back? My patient (the Vietnam vet) described his father as "angry 24/7". That is putting it mildly. My patient also reported his mother daily beat him "on dad's orders" with belts and willow sticks "because you did something we don't know about". He also reported his father forced my patient's brother and sister to beat and hit him while he was tied to chairs with his hands behind him. That sounds like a WWII interrogation method alive and unwell in the United States.

Even worse, my patient alleged that his father raped him anally approximately twice a week from the time he was six to ten years of age. What does that have to do with combat trauma? We're not sure because we never interviewed his father. He was dead by the time I did the evaluation on my patient and he may not have been a reliable reporter given his head wound that obviously damaged his prefrontal lobes. However, any good neurologist or neurophysiologist can tell you that some head injured patients become hypersexual. It's called the Kluver-Bucy

Syndrome (after its discoverers). It can also include hyperactive oral behavior (excessive mouthing of inappropriate objects) but I have not seen that in humans. Beware. You have been warned.

Another head-wounded Vietnam vet would get intoxicated, beat his wife and cut her clothes off with his combat knife.

One more gruesome example of WWII child abuse that goes above and beyond any civilian child abuse I have ever heard of: I had a female patient whose father's ship was sunk in WWII in the Pacific Ocean. To add to the traumas, he had a shipmate trapped under burning rubble that they couldn't get free. It is the sailor's dilemma of which to fear worst, drowning or burning to death. As the flames burned closer and hotter, the shipmate asked her father to kill him so he would die a quick death. The father shot him in the head.

However, this man re-enacted this wartime tragedy for many years after returned stateside. He would take her down to the boiler room in the house in St. Louis, Missouri, stoke the boiler red hot with coals and scream and holler. (I'm not sure what else he did to her but she was a Multiple Personality-now called Dissociative Identity Disorder). In addition to the boiler room, he allegedly made her cook spaghetti for him from the time she was three years old. Unfortunately, one time she dropped the spaghetti on the floor between the stove and the sink. He got so upset he burned her hands in the boiling spaghetti water to punish her.

Again, did this happen to every vet and his children after they came back? No, of course not, but I want to prepare you health care providers for the worst of the worst so that you make sure you give each vet proper 'framing' (of Duty to Warn) regarding child abuse.

The last example that I have is more current. It involves a soldier from Desert Storm. My initial referral was to assess his 22-year-old son (call him K.) for fitness to parent his own child. That child had been taken into state custody after

...

the 22-year-old had a big fight with the child's mother and both of them left the son unattended at a motel. The son had many years of drug abuse, criminal and juvenile acting out and early dropout from high school. However, the 22-year-old reported his father had been shot in the left jaw during Desert Storm. That bullet tore out his right cheekbone, among other things. Another bullet had hit him in the back of the head, causing less obvious cosmetic damage.

Anyway, he said his father slapped him, hit him, kicked him, knocked him down and choked him many times. In fact, his first memory is of his father coming home, slapping him in the face, repeatedly, throwing him across a table and knocking him out because K. had scattered salad on the floor as he was eating it. (This was on K.'s fifth birthday). He awoke to his father crying and saying he was sorry, but he denied doing anything to Kenny to DFS when they came. They came and took K. because his father then pointed a gun at K., said "I'm gonna kill you and put you out of your misery". A policeman outside their window heard his father screaming and entered the house. His father turned the gun toward the policeman and the policeman shot him in the shoulder. K. said he was terrified.

Again, did this happen to every vet and his children after they came back? No, of course not, but I want to prepare you health care providers for the worst of the worst so that you make sure you give each vet proper 'framing' (of Duty to Warn) regarding child abuse.

Behaviors Therapies: Systematic Desensitization
: EMDR (Eye Movement Desensitization and Reprocessing)
: TIR (Traumatic Incident Reduction)
: EFT (Emotional Freedom Technique-also called the "Tapping" technique).
: Prolonged Exposure Therapy (PET)
:Cognitive Behavior Therapy (CBT)

I place Systematic Desensitization and all of the others in the same category simply because they have a similar paradigm. All of them involve repeated exposure parts of the trauma stimuli with the client in control of the exposure intensity and duration. I do not intend to critique EMDR or the others separately. Repeated exposure without recurrence of the full trauma produces **extinction** of the emotional response.

I have heard the army is doing some systemic desensitization using video games (Virtual Vietnam, Virtual Iraq). That may work by showing any combat vet that he is still triggered into high physiological arousal and emotional status by reviewing and acting "as if" he/she were still in combat. Repetition of the video scenes will produce desensitization. It may or may not desensitize everything.

I have also tried to use the "high impact" group therapies. Those involved watching Vietnam War films like "Platoon". Those triggered massive emotional surges for most of the viewers. Sometimes those surges lasted for days. They didn't want to watch them anymore.

Unfortunately, the extinction of systematic desensitization involves repeated exposure. It is <u>critical</u> the exposure be repeated so that the soldiers' emotional surges decline (to both the films and to real-life triggers).

Caveat A: any way this can be approached that is eventually effective will <u>initially</u> trigger incredibly strong, negative feelings.

Caveat B: The troop isn't going to like this version of treatment since it will be so intensely emotional for them <u>at first</u>. It may be so intense as to be counterproductive. The patients and therapists have to decide BUT the patient has to be aware that their arousal will <u>always</u> fade with repetition. Do it fast (five years); do it slow (50 years). Dance your own dance but you'll always pay the fiddler.

Look at the chart below for the ebb and flow of "Triggers".

...

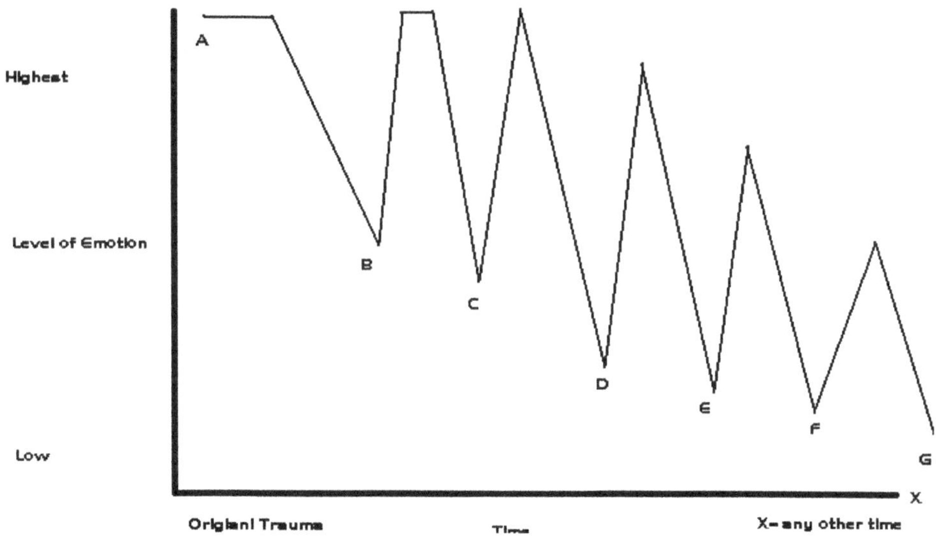

Note that the emotional arousal is the highest and lasts longest at point "A", the original trauma. When the first "trigger" hits at "B", the arousal goes back very high but is not sustained for as long as the original trigger. When the second "trigger" hits at "C", the arousal is high but the surge doesn't last nearly as long as to A or B. Surges at D, E, F and G don't produce as intense or as long emotional surges as the earlier ones produce.

Time "X" (when the emotional responses get bearable) is unpredictable. It can be five years or 50. I saw "bearable" because there are studies that show combat veterans have elevated physiological reactivity to even simple medical checkups 20 years post-Vietnam (Gerardi, et.al, 1994). I'll bet my Korean War vet who was watching "Saving Private Ryan" and was curled up in a ball at the theatre was also having elevated physiological reactivity compared to the other, non-wounded movie viewers That was 50 years post combat!

This is a chart of "extinction" of emotional surging produce by both reality therapy, Classical Conditioning, traditional talking therapy, TIR, EFT, PET and/or EMDR.

This extinction pattern is the same if you chart the emotional response to the same or different triggers of his war experiences. For example, if you repeatedly show the vet the same picture of a tank he was in when it got hit by an IED, his emotional response to it will decrease in the pattern I drew. His emotional responsiveness will also decrease if you show him mixed pictures of the tank, the town it got hit in, samples of IEDs, members of his outfit, and his group banner.

Vets: you are going to do this the hard way (unassisted and unexpected and alone) or the easier way (systematically and intentionally). Trust me. It's going to be tough either way but will save you years of grief if you do it intentionally and with support).

Anger management: This is going to be a primary treatment issue for most of them. This is due to many factors both psychological and neurochemical. The psychological factors are obvious to them: "You pissed me off". The less obvious problem to them is that one of the hormones released during simple stress (cortisol) is also released during full do-or-die. It has been paired in the past with massive physiological arousal and massive threat. It's minor releases during "normal" civilian stresses may put them back in "full metal jacket" for many years.

The literature is rampant with data showing higher anger, domestic violence and abuse of their children. See the additional readings listed at the end of this course for specifics.

Some, but certainly not all, of this violence is specific, unconscious "acting out" what they went through in combat. For example, I had a patient who was a Multiple Personality. Her father had been in the navy in WWII. His ship had been set afire during a battle at sea. There was heat, screaming, yelling and death. A shipmate of his got trapped by some twisted metal and couldn't get out from under it. The ship was sinking and he was burning up from the fire on board. He pleaded for her father to kill him so he didn't drown or burn to death. Her father shot him with a revolver.

...

He came back crazy from the war. He had no one to talk to except, apparently, her. He would take her into their basement, stoke up the coal stove (even in the heat of summer) and be screaming and yelling. He was unconsciously acting out his ordeal on board his ship. He had no one else to talk to so he could only act it out.

She was terrified. She tried to hide. I don't know what else he did to her (except burn her hands in the boiling water when she dropped the spaghetti noodles on the floor-she was three years old at the time and trying to cook for him). She could run but she couldn't hide.

This is a unique example of the effects of the increased aggressiveness of combat vets. The moral to the story: not even their children are safe from their triggered emotional surges.

The children can't help themselves. We adults must help the entire family more.

Chemotherapies: I believe they are absolutely critical IF there has been physical pain and tissue damage, but the medications can't do it all. Research shows a combination of chemicals in the SSRUI category (Prozac, Zoloft, Paxil, etc.) along with anti-convulsants (eg., Neurontin) or atypical antipsychotics work best. The anti-anxieties (anxiolytics) give faster immediate alleviation of the anxiety component of PTSD (but are more obviously addictive). The SSRUIs, atypical-antipsychotics and neuroleptics (anticonvulsants) give help with the emotional lability component of PTSD.

However, I'm not a psychiatrist and don't have prescription privileges, yet. I'm just reporting the current general information cited in Friedman, 2001.

I must warn anyone involved with patients who are psychotropic medications of the following issues that are based on NIMH, NIH and other general researchers.

1) All psychotropic, pain killers and soporifics medications are "addictive" in terms of developing

"tolerance" and need for increased dosage due to decreased therapeutic effect of the same drug administered at the same dosage over a given time. This is due to the tendency of the Central Nervous System (including individual nerve cells, neurons) to decrease responsiveness to the same stimulus at the same intensity, not matter
what that stimulus is, be it chemical, visual, auditory, tactile or visual. That reduced responsiveness to constant stimulus is called "adaptation" at the receptor level and "habituation" at the central nervous system.

2) The SSRUIs take four to six weeks to reach therapeutic level. If your patient reports therapeutic effect before that minimal four weeks, attribute the effect to Mr. Placebo.
3) Many patients I know quit taking the slow-building psychotropics like the SSRUIs because they don't notice an effect soon enough (even if they are told about the 4-6 week buildup) and quit taking them because they want IMMEDIATE results. That difficulty can be easily dealt with giving the patient immediate doses of faster acting anxiolytics (like Valium) <u>until</u> the slower-acting SSRUI takes effect. Then the valium can be tapered off and used prn for more severe anxiety attacks.
4) Many people quit taking their SSRUIs or antipsychotics "cold turkey" because they don't feel like they need it any more. This reduced emotional lability, for the severely physically traumatized, turns out to be mostly due to the effectiveness of the SSRUI and the patient goes on a "manic" drug withdrawal rebound and often has to be hospitalized (or incarcerated).
5) The first time you patient talks about going off an SSRUI, have them talk to the prescribing physician and see if that physician will taper the patient off the medication. If that physician refuses to withdraw the patient or advise them how to taper, I

...

suggest you keep quiet unless you are a licensed medical doctor or a registered pharmacist.

6) Beware the "atypical antipsychotics", among them being Zyprexa, Risperdal, Abilify, Seroquel, et al. These are second generation supposed anti-psychotic medications that also have FDA approval to use for Bipolar. However, I have seen them heavily prescribed for children and adolescents with excessive anger. They have also been heavily prescribed to "manage" patients in nursing homes. There are several problems with this entire category of psychotropics. First, they are 'associated with the onset of adolescent diabetes. They do this by elevating blood sugar levels in ways I care not to discuss. Secondly, long-term use of them has recently been reported to be "associated with" a five times higher risk for sudden cardiac failure. See http://www.naturalnews.com/025724.html, among many.

Hypnosis-don't bother with the newly returned vet. He or she is too raw and probably will not closer their eyes because they reduce their external information.

B) Family therapy-This can be used to educate the veterans and their families about the soldier's issues of war and all of their issues about readjustment that I've talked about in this manuscript. Pray to God you don't have a "GI Jodie" issue People get killed over that

C) Group therapies-these can be unstructured or unstructured. Most members seem to like structure. As I mentioned before, they can be low impact (talking about things) or high impact (viewing combat footage or movies). I repeat the caveat about the Emotional
Tsunamis any high intense approach will initially take.

It is crucial the vet be told before any high impact procedure are conducted, that they will have high emotional responses…at first, but that their emotional reactions will eventually fade away.

Danger to Self and Others

A) Suicide prevention: Danger to Self

Question: How many soldiers really suicided after they came back from Vietnam?

Answer: Nobody really knows. Anti-war groups said twice as many Vietnam veterans suicided after they got back as got killed during the war. One website said that 20,000 suicided from the end of the war to 1993. However, one physician on that same website estimated up to 200,000.

He reasonably explained the discrepancy by stating, "the reason the official suicide statistics were so much lower was that in many cases the suicides were documented as accidents, primarily single-car drunk driving accidents and self- inflicted gunshot wounds that were not accompanied by a suicide note or statement."

("Suicide Wall, Suicide statistics" at http://www.suicidewall.com/SWStats.html)

There is also another phenomenon called "Suicide By Cop". This is not unique to veterans but it is another way to self-destruct without getting it labeled a suicide. This involves getting into a shootout with policeman without taking cover. The person often gets the attention of several policemen by doing something outrageous, then begins shooting at them. They, of course, return fire after going through their less lethal maneuvers which the person, of course, disobeys. The police, in self-defense and as the last resort, shoot the person dead. It's not an overt suicide but it's obviously a self-destruction.

Question: How many soldiers from OIF/OEF will suicide?

...

Answer: Too many already.

We can be all clinical but “rob D27” on the msn PTSD chat group wrote: “I am ex-Green Beret and I think I have lost the ability to care, physically and emotionally. There was a time last week when I almost ended it all, I don't think my wife understands and I don't know how to explain it to her, I know she loves me, but I don't love myself. I don't know where this is going or why I am on this site, but I feel like there is more, I don't feel I fill my obligation as a citizen. I have a plastic shin, plastic hip, crushed skull and a crushed spine, also just had carthledge (sic) injections in my wrists and my knee is about to blow, they can't get my white blood cell count right. What happens when there are no more missions or long deployments? I spent 12 years of my life in the Army for them to throw me out I was a Senior NCO, the only thing I knew was my job, now I can't even do that, so I am a scab on society. It is funny it took two ratings to get 100% from the VA, and I can't get SSD, I guess because I am a citizen, not an illegal alien, why did I waste my time if this is the way it ends. Guess, what my Phsyc. (sic) was for SSD at my appt. Iraqi, told me I was fine and my injuries don't effect my daily living, what an idiot, he is just mad because I probably killed his whole family, I wish he would have been there too. It makes me cry to know that are (sic) country is being overrun by foreigners, so deserving soldiers who have spent years in Military paying taxes can't get SSD, because the crap they pay me is nowhere near my salary in the ARMY, I did not ask to get out! So as I watch my bills fall behind, and if something does not change, I will either be on a street corner, or dead, and I refuse to beg I am to honorable. I am tired of the nightmares the flashbacks the headaches the whatever hell else happens during a normal day. America does not care about the Military, and a hand full care about the soldiers and the only reason they care about the soldiers now is because they are in Iraq. I have done every school the Army would let me, Sniper, HALO, Ranger, Special

Forces, Master Fitness, Pathfinder, Combat Diver, blah blahh..... it doesn't matter. Oh yea! God Bless America."

Will anyone help him? When someone is blown up this bad, he needs a lot of help. This is their reality after the band goes home and the television interviews are over.

The VA and DOD have repeatedly announced that discharged veterans are suiciding because of marital, financial and interpersonal stress. That is a minimization of the combat-trauma related emotional surges they are having. Not everyone suicides due to marital stress. Not everyone suicides due to financial stress. Not everyone suicides due to interpersonal stress. Suicides are done to escape enormous emotional pain when the person feels they have no other way out of the pain!

My repeated responses to more information have been repeatedly blocked or totally ignored.

New Orleans Veterans Administration hospital data (2005) suggested the following factors were related to combat veteran suicides:

- Group/unit cohesiveness
- Rated leadership preparation/ability
- Untrained for activities (cooks, transport people caught in line of fire)
- Post Discharge stresses (benefits, insurance, VA Hospital struggles, financial, and relationship)

That data doesn't mention the Dear John cell phone calls and Dear John text messages the OEF/OIF veteran in combat can get that has devastated combat vets ever since I know of. It also minimizes the inescapable (they think) agony all suiciders have leading them to suicide. For example, any reader of this book has had benefits, insurance, health care, financial and/or relationship stresses. However, I guarantee you that none of you have

...

successfully suicided (or else you wouldn't be reading this book) and most people with these stresses never try suicide.

The Veteran's Administration's Toll-free suicide hotline number is: 1-800-273-TALK (8255)

Gregg Zoroya, of USA Today, reported:

- Nearly 40% of Army suicides in 2006 and 2007 were taking psychotropic drugs like Zoloft and Prozac for depression and PTSD.

- Nearly 60% of 948 Army suicide attempts in 2006 had been seen by mental health providers before the attempt - 36 percent within just 30 days of the event.

- More than 43,000 U.S. troops since 2003 were sent into combat even though they had been listed as medically unfit in the weeks before their scheduled deployment.

- The "typical" soldier who commits suicide is a member of an infantry unit who uses a firearm to carry out the act, according to the Army.

- 53% of veteran suicides from 2001-2005 came from the Guard or Reserve population; for a period during 2005, they accounted for about 50% of forces serving in Iraq and Afghanistan. However, when averaging all war years, they made up 28% of all U.S. military forces deployed. 100,000 OEF/OIF vets have sought help for mental health issues, including 52,000 for post-traumatic stress disorder alone.

- According to the DoD, there were almost 2,200 active-duty soldier suicides between 1995-2007.

- CBS News reported in November that there were at least 6,256 veteran (of all wars) suicides in 2005 [this figure includes data collected from 45 states; the figure is, therefore, higher if taking all 50 states into account]. That's 120 each and every week. In addition, on any given night, nearly 200,000 veterans are counted among the homeless.

- In 2005, OEF/OIF veterans aged 20 through 24 had the highest suicide rate among all vets, about 2-4 times higher than their civilians peers. (Civilian suicide rate: 8.3 per 100,000; Veterans suicide rate: between 22.9 and 31.9 per 100,000.)

- A 2007 survey of U.S. troops revealed that about 12% of OIF and 17% of OEF combat troops are taking prescription antidepressants or sleeping pills to help them cope.

- The new VA suicide prevention hotline, 1-800-273-TALK (8255), recently reported that it's received more than 55,000 calls, averaging 120 per day, with about 22,000 callers saying they were veterans.

(from http://ptsdcombat.blogspot.com/2008/09/oefoif-veteran-suicide-toll-15-of.html)

The current body count from too many sources to cite are that 1 veteran a day who is still in-service is suiciding and 22 a day who are discharged are suiciding, especially those over 55 (i.e., Vietnam veterans).

Many sources also report that there have been more American soldiers suicide in every year of the OEF/OIF activity than were killed by the enemy in that year.

B) Danger to Others: I've seen little data on the other issue of dangerousness…danger or harm to others. Minimally, this involves increased likelihood for domestic and child

…

abuse (see Taft, et al, 2007). At the extreme, dangerousness to others involves murder.

I must repeat again, that this dangerousness to self or others is highest in the troops with brain damage from either penetrated or closed-head trauma!!!!!!!!!!

Traumatic Brain Injury (TBI)

The effects of brain trauma are very deceptive to the average civilian. They are one of the least understood injuries of the medical profession. Unfortunately, they are being called the signature injure of the OIF/OEF theatre due to the roadside bombs, IEDs, and copper shaped- charges devised by the insurgents.

The systematic neurological study of head wounds only dates back to 1848 when a 25-year-old American railroad worker named Phinneas Gage accidentally blew a six-foot, dynamite-tamping bar through the front of his cheek and out the top of his head in the front of his skull. A picture of his real skull is below on the right. A mock-up is on the left:

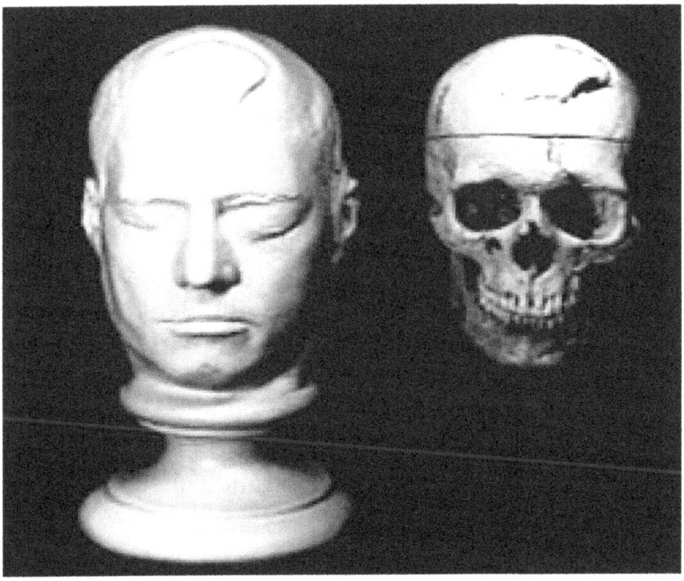

He was leaning over the hold the dynamite was in, tamping it with the steel rod. The tamping ignited the dynamite. Note the insertion point of the bar under his right cheek and exit point above his left prefrontal lobe.

...

The steel rod damaged the pre-frontal lobes of his brain. He was completely lucky it didn't kill him.

He miraculously physically survived the blast and lived many years afterward, but he was a changed man. He was once described as "a dependable and likeable crew boss" but had become "an irresponsible and rowdy ruffian". (Damasio, 1994). To be more specific: before the blast, he was a calm, civil man. After the blast, he was an emotionally volatile (moody) man. When he was happy, he was ecstatic. When he was unhappy, he was morose. When he was sad, he was inconsolable.

His emotions had become greater and more extreme.

His vision was unchanged. The brain under the back part of the skull process vision and it had been uninjured.

His hearing was unchanged. The brain under the skull under the ears that Processes hearing and they had been uninjured.

He could taste as usual. He could smell as usual. He could move normally.

All of those brain parts were uninjured so those functions were unchanged. He appeared basically unchanged.

The only parts of his brain that were injured were the parts that controls the cognitive (mental) "Executive Functions" (the prefrontal lobes) and emotions (by damping them down), the orbital-frontal lobe.

He had become an unreasonable, erratic, uncontrolled ball of affect (emotion).

This is what has happened to some of our troops in past and present wars. Our understanding of brain physiology (function) was very rudimentary in wars before Vietnam. We are now at the apex of our understanding of neurophysiology. There is no need to ignorant of the effects of head injury on personality and self-control. Unfortunately, it remains high in most civilian and military members.

So what are those cognitive (mental) "Executive Functions"? In 2014, I wrote:
"The mental, cognitive processes that occur when these association cortexes are making mental associations are labeled by many (Pribram, 1973; Lhrmitte et al, 1972; Lezak; 1982; Shallice, 1982) as "executive functions". A fairly comprehensive and lengthy list of the generally accepted executive functions includes sustained Attention and Perception, Intentional Mobility (movement), Memory, Planning, Intelligence, Temporal Integration, and Language. If one looks carefully, one will see these are all of the cognitive processes we humans assume to be higher intelligence.
Some of these were subdivided by Furster, 1997 (p. 152-156), into several other subsections: He divided Attention/Perceptions into: attention (both low alertness and sensory neglect), external interference (distractibility), disorders of visual search and gaze control, difficulty sustaining attention, difficulty sustaining attention from internal interference (usually called "poor impulse control" or hyper-distractibility in clinical cases), and defective motor attention set (attention deficit). He subdivided Mobility into hypokinesia (lack of spontaneous motor activity) and hyperkinesia (excessive and aimless movement). This author will not list all of Furster's possible categories because I feel many are overlapping and/or irrelevant to the thesis of this book. However, the list so far does include the foundation for higher cognitive activities: thinking, abstraction, critical analysis, judgment and impulse control.
The subtly of the effect of the injury (either obvious or subtle) on these association area's abilities is deceptive. What is lacking with the prefrontal-lobe or parietal-lobe injured? What don't they have? What can't they do? What are they missing? What are they lacking? It is extremely obvious what they can do. They can see as well as any of us. They can hear as well. They can walk. They can talk. They can believe. They can want. They can feel. They think they can think. We think they can think.
So what can't they do? They can't stop!

...

What this author would like the reader to appreciate is that all of the higher mental activities labeled "executive functions" require not doing things: stopping (not going), quitting (not doing any more), calming oneself (not flying off the handle, not continuing a rage for hours at a time), being reasonable (not being irrational or emotionally-driven), comparing alternatives and options before one acts (thinking before acting). The subtly brain-injured person often appears normal unless they have obvious external sequellae (after-effects) such as scars, indentations, lost appendages, and/or lost sensory organs. However, they can't stop, can't quit, can't calm themselves, can't reason, can't be reasonable and can't truly think (compare, critique, connect the dots). They typically run on emotional surges rather than thought-out, preplanned activities (hyperkinesia). They emotionally over-react because the brake lines to the brakes (the orbital-frontal lobe) has been severed, leading the amygdala to function with excessive, un-dampened excitation."

The brain injured person can't maintain thoughts in their heads like they used to, they can't evaluate them (think) like they used to, they can't calm themselves any more.

So how many TBI soldiers do we have? Melissa Healey from the *Los Angeles Times* reported that 9000 troops have been evaluated for TBI through January 2009. Unfortunately, a Rand Corp. study estimates that as least 180,000 combat troops have sustained head wounds capable of causing brain injury with a realistic maximum number of 360,000 exposed. (Reported in http://ptsdcombat.blogspot.com/2009/10/combat-clips-selection-of-oefoif.html, October 30, 2009)

I have repeatedly evaluated children with subtle epilepsy or other brain damage who have focal spots that impair self-control. They are initially, erroneously diagnosed as ADHD (attention deficit, hyperactive disorder) because they have low self-control.

I have evaluated two adult men who have caused serious injury or death to their children by shaking them. Both of them had undiagnosed brain injuries that had severely diminished their self-control. The most common factor among convicted murders is brain damage. It goes undetected and is unappreciated (for its potential for lethal violence).

Not all head injuries cause murderers. Not all head injuries produce severe lack of self-control. However, the sophisticated neurological screening devices we now have (MRIs, functional MRIs, PET-scans, MEGs) should provide the head-injured veteran and their family more scientifically-based evaluation of the consequences of their head injuries AND indicate prudent and state-of-the-art interventions (both short-term and long, long, long-term). There are also many less-costly, but effective, paper-and-pencil tests that accurately measure some brain-dysfunctions related to head injury. These include complex intelligence tests like the Wechsler Adult Intelligence Test (WAIS), simple intelligence tests like the Kaufman Brief Intelligence Test (KBIT) and the Bender-Gestalt Visual Motor test. The military is using others but I do not remember which ones.

Initial (but frequently replicated) research on murder (D. T. Lunde, 1975; Pitman, Orr, Forgue, de Jong, & Claiborn, 1987) found that most murders occur on weekends, most of them on Saturday night between 8 pm and 2 am, peak in July and December, occur more frequently in Southern states (44% of all murders) and involve, on average, a 20-year-old male killing another less-than-30-year-old male with a gun. 90% of the time, the victim and the perpetrator know each other.

This typically adds up mostly to be a brain-damaged friend killing another friend or enemy during a late-night drinking/drugging "party" on a holiday.

Can we prevent all injuries to self or others? No. There are too many situational (in the current time) factors that trigger or defuse the emotional surges behind suicide and homicide. The triggers are all situational. Someone says the wrong thing; someone does the wrong thing. The defusers are also situational. Someone talks them out of it; someone does a right act to calm them. Sometimes it's just sheer luck.

...

For example, I had one Vietnam combat veteran who was in a rage after finding out his wife had been recently unfaithful to him, 20 years after he had come back from Vietnam. He was going out the door with a loaded rifle to kill his wife's lover when he accidentally knocked over a flowerpot with the barrel of his rifle as he turned to shut the front door. He stooped down and cleaned up the flower, pot and soil. His rage had subsided by the time he got the mess cleaned up. He was unemotional enough to unload the weapon, go back inside the house and cry. The flower pot would not have been knocked over if he had been carrying a rifle. The flower pot would not have been knocked over if it had been winter time. So close to homicide; so luckily, everyone stayed alive.

There are also many more restrictions (patient's rights) to institutionalization. I believe those liberal patient's rights have led to many unnecessary injuries to self and others. Many families of deceased patients also wish there were more authority to institutionalize those needing high supervision than they got.

Descher, et al (2003) found the combined rate of suicide, homicide, or suicide-by-police for previously hospitalized PTSD vets was 13.8%. That's 13.8% of all hospitalized vets, not just those diagnosed with PTSD. Is 13.8% good or bad? Is 13.8% enough, too much? I don't know. However, I'll bet it's improvable.

I also want you to keep in mind that our current, middle-eastern troops have been in combat situations for as long as twelve_years (as of 2017), in an arena mixed with civilians and combatants, adults and children, men and women. The mixed arena is the same as the Vietnam combatant faced, but he often was home after one tour (12 or 13 months).

I believe the extended fighting and mixed arena nature of the current middle-eastern conflicts will increase the mental and physical damage to our troops compared to prior wars.
While TBI is commonly attributed by the American military to IEDs and "indirect fire" (mortars and artillery), there are many,

many other sources of head trauma in the military that few other authors report. I do so here with the following list:

a) Falling off vehicles, walls, etc. These falls can be especially damaging because the soldier often has his ruck sack or other heavy items in their hands when they fall and the American combat vehicles are now six to 10 feet tall.
b) bar fights (fists, yes; very thick beer bottles)
c) car/motorcycles wrecks
d) getting electrocuted
e) falling down stairs
f) having heavy objects fall on them (most often seems to be .50 caliber machine guns and barrels)
g) getting thrown into ship walls during rough seas
h) getting knocked over a cliff
i) getting shot in the helmet by a sniper (and living)
j) Neurotoxins like Mefloquine/Lariam

The last factor (Mefloquine/Lariam) is a controversial issue. The drug was used as an anti-malarial drug supposedly superior to sulfa. It was first produced in the 1970s and the articles I have found say it was last issued the by American military in 1993. It was also issued to British and Australian soldiers. It is alleged by many to produce extreme emotional lability and perceptual disturbances. There is a Facebook group titled "Veterans Against Mefloquine". It is a closed group but those veterans with relevant concerns should feel free to apply for membership. However, I must warn all concerned with this issue to read the following chart published on the internet. I find it excellent to separate the overlapping effects of PTSD, TBI and Mefloquine. I cannot find the original source so I cannot give credit to the author. I just assure the reader I did not develop it. I also refer to reader to research done by Remington Nevin, M.D., and his group, easily accessed on the Internet by searching by his name.

...

ISSUES Ver 1.4	MEF	PTSD	TBI
ANGER	X	X	X
DEPRESSION	X	X	X
ANXIETY	X	X	X
UNUSUAL BEHAVIOR	X	X	X
MOOD CHANGES	X	X	X
SLEEP PROBLEMS	X	X	X
CONCENTRATION	X	X	X
CONFUSION	X	X	X
IMPULSIVENESS	X	X	X
PERSONALITY CHANGE	X	X	
AGITATION	X	X	
NIGHTMARES	X	X	
FLASHBACKS	X	X	
FEAR OF CROWDS	X	X	
PARANOIA	X	X	
RESTLESSNESS	X	X	
SUICIDE IDEATION	X	X	
HYPER ALERTNESS	X	X	
UNABLE TO MULITASK	X		X
MEMORY LOSS	X		X
LIGHT SENSITIVITY	X		X
DIZZINESS	X		X
NAUSEA	X		X
HEADACHES	X		X
VERTIGO/BALANCE	X		X
SPEECH (APHASIA)	X		X
RINGING IN EARS	X		X
VISUAL IMPAIRMENT	X		X
TREMORS / JOLTS	X		
PRURITUS	X		
SWALLOWING	X		
PNEUMONITIS	X		
MUSCLE WEAKNESS	X		
IRREGULAR HEARTBEAT	X		

There is little research that realizes and separates these three factors out and there is plenty of anecdotal stories about the bizarre perceptual experiences veterans (and civilians) have had after taking this drug. I have two cases of veterans who had taken Lariam and later had bizarre visual experiences. One involved a veteran who was riding his motorcycle along and his visual field invert 180 degrees. In other words, his visual field turned upside-down. Top became bottom and bottom became top. He was so alarmed he had to pull his motorcycle off the road. He did say his visual field righted after a brief passage of time. It never happened again.

The other case involved an OEF/OIF veteran. He was driving his car along and visually experienced driving into the road. Not along the road. Down into the road. He had to pull over too. This visual aberration also passed very quickly.

I have not heard of it producing any other sensory distortions.

I must note my own observation that while both PTSD and TBI produces headaches, TBI produces a unique headache that is tied to photosensitivity that PTSD does not produce. People with TBI -related headaches have to wear sunglasses inside and out, unless it is extremely dim lighting. The other TBI unique mental change is documented in the TBI neuropsychological test data: they think accurately but more slowly. This is measured by some WAIS subscales that are timed (Symbol Search and Coding). It is also picked up by the Ruff 2s and 7s test.

The mental, cognitive processes that occur when these association cortexes are making mental associations are labeled by many (Pribram, 1973; Lhrmitte et al, 1972; Lezak; 1982; Shallice, 1982) as "executive functions". A fairly comprehensive and lengthy list of the generally accepted executive functions includes sustained Attention and Perception, Intentional Mobility (movement), Memory, Planning, Intelligence, Temporal Integration, and Language. If one looks carefully, one will see these are all of the cognitive processes we humans assume to be higher intelligence.

...

Some of these were subdivided by Furster, 1997 (p. 152-156), into several other subsections: He divided Attention/Perceptions into: attention (both low alertness and sensory neglect), external interference (distractibility), disorders of visual search and gaze control, difficulty sustaining attention, difficulty sustaining attention from internal interference (usually called "poor impulse control" or hyper-distractibility in clinical cases), and defective motor attention set (attention deficit). He subdivided Mobility into hypokinesia (lack of spontaneous motor activity) and hyperkinesia (excessive and aimless movement). This author will not list all of Furster's possible categories because I feel many are overlapping and/or irrelevant to the thesis of this book. However, the list so far does include the foundation for higher cognitive activities: thinking, abstraction, critical analysis, judgment and impulse control.

The subtly of the effect of the injury (either obvious or subtle) on these association area's abilities is deceptive. What is lacking with the prefrontal-lobe or parietal-lobe injured? What don't they have? What can't they do? What are they missing? What are they lacking? It is extremely obvious what they can do. They can see as well as any of us. They can hear as well. They can walk. They can talk. They can believe. They can want. They can feel. They think they can think. We think they can think.

So what can't they do? They can't stop!

What this author would like the reader to appreciate is that all of the higher mental activities labeled "executive functions" require not doing things: stopping (not going), quitting (not doing anymore), calming oneself (not flying off the handle, not continuing a rage for hours at a time), being reasonable (not being irrational or emotionally-driven), comparing alternatives and options before one acts (thinking before acting). The subtly brain- injured person often appears normal unless they have obvious external sequellae (after-effects) such as scars, indentations, lost appendages, and/or lost sensory organs. However, they can't stop, can't quit, can't calm themselves, can't reason, can't be reasonable and can't truly think (compare, critique, connect the dots). They typically run on emotional surges rather than thought-out, preplanned activities (hyperkinesia). They emotionally over-react because the brake lines to the brakes (the prefrontal lobes) have been severed or overwhelmed with excessive, undampened excitation.

...

THE OTHER INVISIBLE WOUNDS

There are two other sources of death and chronic disease that veterans have been exposed to that have recently come to light, sleep apnea and the burn pits.

Sleep Apnea

Is defined by the Mayo Clinic as "a potentially serious sleep disorder in which breathing repeatedly stops and starts. If you snore loudly and feel tired even after a full night's sleep, you might have sleep apnea. (**https://www.mayoclinic.org/diseases-conditions/sleep-apnea/symptoms-causes/syc-20377631**). Note that one does not snore breathing out. One snores while breathing in. That means one is not getting normal inhaled oxygen during sleep.

The causes of it are many, the most common one as mentioned by that Mayo Clinic article is obesity and given that 1/3 of all Americans are "morbidly obese" and obesity is now a world-wide problem so sleep apnea is affecting people world-wide at an ever increasing rate whether they are aware of it or not.

Those with sleep apnea can suffer day-time symptoms of depression, excessive fatigue, sleepiness and even falling asleep while working, driving, watching television or other passive activities.

While snoring is the most obvious symptom, one can snore but not have significantly reduced oxygen and one can not snore but have sleep apnea. Other dysfunctional signs of sleep apnea include simply stopping breathing while sleeping or waking up gasping for breath (which many sleepers mistake as awakening from a nightmare). Asthma, COPD and other respiratory disorders can also produce sleep apnea and reduced oxygenation.

Less obvious factors that can produce sleep apnea include the mental problems of PTSD, high anxiety (whether diagnosed as Generalized Anxiety Disorder, Obsessive Compulsive Disorder),

Major Depressive Disorder and Bipolar Disorder (due to the actual symptoms of the person, not just the global diagnosis).

Even less obvious factors that can produce sleep apnea include GERD, chronic physical pain in upper and/or lower body, tinnitus. The pathophysiology of all of these processes is a multifactorial, multidimensional dance between mind and body.

The seriousness of sleep apnea is severely underrated. It is documented to result in death due to many factors, none of which involve directly dying in your sleep from lack of oxygen. The hypoxia (lack of oxygen) produced by sleep apnea can damage heart muscles, heart nerves and heart valves. Any dysfunction of any of them can produce pooling of blood in the lower heart chambers (ventricles) and blood pooling clots. The clot(s) eventually get pumped out of the ventricles, move into the ascending and/or descending arteries and can lodge in brain, lungs or limbs, resulting in death or ischemia (cell death due to lack of oxygen).

If that isn't bad enough, sleep apnea is "associated with" (we can't say "cause" due to other physical factors being associated with the problems, too) of lupus erythematosus, multiple sclerosis, diabetes mellitus 2, fibromyalgia, chronic fatigue syndrome, hypertension, migraine headaches, rheumatoid arthritis, vertigo (aka Benign Paroxysmal Positional Vertigo). I am fully aware that physical medical professionals view many of these as "autoimmune" disorders and look for physical markers. However, their relationship to psychological processes makes them psychosomatic disorders, not just purely physical disorders. I am also fully aware that physical-medicine oriented people view psychosomatic medicine with disdain and disrespect.

The Invisible Wound from the Invisible Wound

There are several mechanisms for treatment of sleep apnea including the CPAP/BIPAP/AutoPAP machines, the "bongo" (a pair of nose plugs, the Mandibular Adjustment Device (MAD), Inspire implant, and various surgeries (Radiofrequency volumetric tissue reduction, Uvulopalatopharyngoplasty,

Maxillomandibular advancement, Anterior inferior mandibular osteotomy, Genioglossus advancement, Midline glossectomy and base of tongue reduction, Lingual tonsillectomy, Septoplasty and turbinate reduction, Hypoglossal nerve stimulator, and Hyoid suspension (**https://www.sleepfoundation.org/sleep-apnea/surgery-for-sleep-apnea**).
Of course, most of those are invasive, carry risk for infection, some have prolonged recovery times, and none don't always succeed in stopping sleep apnea.

If all of this isn't bad enough, there is one original version of a CPAP machine brand (Phillips) that had a design flaw that melted plastic or foam inside the machine and allowed the fumes from the melted material to enter the air system of the machine and fed into the respiratory system of the breather-producing a mini-burn pit in their lungs every time they used it. That defective system has been nationally recalled but I still at the time of this latest writing (2023) hear of Veterans that did not get that recall notice and are still using that version.
The most common symptoms I have learned from users of that faulty machine are rashes around the mask and excessive salivation, sometimes choking the patient in their sleep.

The Burn Pits

A comical, miniature version of the burn pits is well known to most veterans, stirring the diesel fuel-feces mixture doing "latrine duty". Yes, the concoction smells horrible and is a form of punishment, hence degrading, but in the age of static military bases and high tech components (especially in the wars in Afghanistan and Iraq), there were massive trash piles created in hostile lands that were burned, usually in open pits. They weren't supposed to be handled this way because the pentagon awarded contracts to companies like Kellog, Brown and Root to set up incinerators to burn trash in enclosed, environmentally safe machines. However, few of those were actually activated and the ones that were placed failed sometimes but were never serviced. This led to open air trash pits spewing toxic fumes and smoke for months at a time with service men and women being exposed to those fumes on a daily basis. Not surprisingly, this has produced severe respiratory problems for thousands of veterans from those deployments. The exposures were so unhealthy that the Veteran's Benefits Administration now awards "presumptive" compensation for a variety of respiratory problems simply if the Veteran was in the area of those burn pits (but the award is now limited automatically only IF it is claimed within ten years of exposure-which rarely happens). That window of opportunity may be extended as further research shows the full extent of the effects of those burn pits on Veterans

THE GUILTS

There are five different guilts I know of concerning combat veterans: 1) Survivor guilt, 2) participant guilt, 3) atrocity guilt, 4) friendly fire guilt, 5) nonparticipant guilt, 6) compensation claim guilt. A "joke" we had from the Vietnam War was:
"Everyone has guilt about it except the ones that got killed".

Unfortunately, I see less **nonparticipant** guilt from this current war (OEF/OIF) than Vietnam, probably because our current OEF/OIF troops are real and sorta volunteers. There is certainly a much smaller percentage of able-bodied adults fighting the current wars (2%) than there were during WWII, Korea or Vietnam when almost-everyone was drafted.

I have already talked some about **survivor** guilt, especially in the clip and transcript about SSgt. Lingo during the CVT3 segment.

Participant guilt-although I know that all combat troops take the attitude of "kill or be killed", "better them than me", I still have reports from many men that say they felt bad, felt guilty and were emotionally effected by the killing when they actually heard or saw the effects of their weapon on their target. I may have written earlier about the Vietnam veteran in the Mekong Delta who actually saw the effect of his attempt to shoot to wound a fleeing Viet Cong soldier. He shot at his leg. His bullet hit his leg. He expected the Viet Cong to fall to the ground, wounded. His bullet actually hit the Viet Cong in the bone of his

...

leg. That produced much more impact on the Viet Cong man. The total power of the bullet hitting his leg bone caused the shot to whip the man's entire leg around in front of him. It happened to pinwheel his leg over, around and into his head. His own heel smashed into his skull, killing him. Our kind shoot-to-wound solider felt horribly bad seeing his victim get killed in such a ghastly, unexpected way.

Another example I heard of: An American Korean War soldier's weapon was a flamethrower. He burned some enemy soldiers to death. After he came back to the States, he began hearing voices telling him "they" were going to burn him. He came back from the war and immediately starting cutting his own face (para-suicidal behavior). He began drinking alcohol heavily. Why did he start cutting himself? Why did he begin drinking heavily? Parasuicidal (self-harm) behaviors are usually done to inflict physical pain so they don't feel emotional pain as badly. Why did he start drinking heavily? To numb himself from the bad emotions he felt that the cutting didn't always end. Why did he cut his face? I don't know but I suggest it was because he felt disgraced or dishonored (had "lost face") by what he did with the flame thrower. Why would he feel so bad about the effects of the flame thrower? I suggest it is because 1) it is a short-range weapon and he probably heard, saw and smelled the effect his weapon had on his enemy troops. Have you ever heard the screams a horrible injured human makes? Ask those who have. They will remember it forever.

Atrocity Guilt-not many combat vets admit it but the Mi Lai massacre was not the only atrocity that ever happened to civilians in wartime. Most participants won't talk about it out of fear of court martial. I have never worked with or obtained any information from a solider that did participate in a civilian atrocity so I cannot help anyone on this issue.

Friendly-fire guilt: there is a "fog of war". In the pre-gunpowder days, it was caused by the massive clash of bodies and the literal fog of black powder. They actually had drummers and buglers to let each side know where their side of the line actually was. In the more current wars, the pace of the war, the lack of stable lines (our side/their side), the flurry of hand-to-hand combat and the night time nature of much war makes clear visibility absent at times. At that point, it's fight by pure reflex, shoot first, ask questions later. Our most contemporary, well-publicized victim of friendly fire was Pat Tillman, a professional football player for the Arizona Cardinals who resigned from the football team and enlisted in the military after 9/11/2001. He died in a friendly-fire situation soon after he got in Afghanistan in 2004.

Compensation Claim Guilt: I have heard that many Veterans avoid applying for compensation because they think or say, "I know people who had it worse than me", and "I don't want to take the money from people who deserve it more than I. That is an injustice to them because of all the Invisible Wounds that combat can produce and even though they are "invisible", they are still lethal, degrading, libido-killing, quality-of-life killing for the rest of their life. I have a morbid saying that might offend some but here it is:

"In war, the lucky ones only get killed."

The "stigma" of therapy? I have seen much in the public print, especially from military sources (e.g., The Veterans' Administration Center for Post Traumatic Stress Disorder, Iraq War Clinical Guides), that combat vets won't reach out for counseling or suicide prevention due to the "stigma" or negative consequences of receiving help. I concur that there might be more reluctance to seek mental health help by troops in the current volunteer army that view the army as a career. However, in my 43 years, I never seen a seriously suicidal patient who resisted seeking help at some point in their decline. It may come early in the progress to the final (sometimes fatal) resolution. It may come later. However, I have always seen they are seriously seeking a non-fatal resolution to their real agonies and are amenable to sincere assistance. I am aware that many veterans reach out to other veterans in Facebook groups such as Veterans 2 Veterans Info.

I was shocked to find out that many veterans facilities do not have any organized crisis support. I was more shocked to find

...

out they do not offer services besides eight-to-five, Monday-through-Friday. Remember, most murders take place on weekends and in the wee hours of the night.

I was shocked to find out it is common to have two month waiting lists for VA counseling. As usual, we can find all the money asked for the war, but little for the warriors. I guess "you go to war with the army you have---not the army you might want or wish to have at a later time." (Donald Rumsfeld, Secretary of Defense, December, 2004 Town Hall meeting).

It appears to me that the DoDs lack of respect for the mental health and perceived (or real) lack of understanding from the mental health arm of the Veterans Administration (VA) is an extremely serious impediment to the returning combat veterans trusting (and therefore reaching out to) VA mental health support.

Stigma? Hey, come on! I know a psychologist who was a conscious objector during the Vietnam war but many combat vets have opened up to him for decades…most of the time even knowing what his choice was for Viet Nam.

Stigma? What stigma? Unwillingness to open up? What unwillingness? I've never seen ANY unwillingness for a combat vet to open up if they felt respected and not condemned.

Stigma? There are ways around them. Many ways. Many, many ways.

Theoretical Considerations

I have two general hypotheses about PTSD and the effects of combat trauma. Both of them directly impact the diagnosis, treatment and case management of combat veterans. The first hypothesis postulates a tri-phase nature of PTSD evolution and, therefore, treatment. I am generalizing this hypothesis from the combat and civilian trauma population I have worked with.

This hypothesis is a work-in-progress so don't hold me to it as if it is currently set in concrete.

Also don't try to hold me to it because I am well aware of the criticisms of any and all categorization or stage theories. Those criticisms and objections include a) the true number of any distinguishable and reliable subcategories for anything is probably seven (take any of my college classes to find out why), b) not everyone will go through all three (or seven or however many stages), c) people will regress to "lower" stages and progress to "higher" stages in uneven steps, d) some people may never complete (completely "work thorough") some issues. Look at the personality theories of Eric Erickson or Elizabeth Kubler-Ross' Death and Dying for parallel criticisms of their stage theories.

Hypotheses 1: The Three Phases of PTSD

1) "In the trenches", "At the front", "In the boonies", "In the bushes", "In the sand pile", "Outside the wire" (or whatever slangs the troops in that theater use for still being "in harms way".

Nobody knows or talks much about being traumatized while they are still in the fight because everyone is still busy trying to keep from getting killed. Part of the syndrome, at that point, is a survival mechanism (hypervigilance, not taking about it, using avoidance, sleep disorders) relying on combat training and instincts instead of processing the death and killing, loading up with alcohol, sex and rock 'n roll on R' n R').

2) Discharged with "brass balls". Everybody luxuriates in the thrill of being back home at first. The mental focus is on celebrating the survived soldier, no matter what shape he is in. He can have his legs blown off, his jaw shot off or anything else shattered and tattered. He/she made it! That's the most important thing. Right? Right! Unfortunately, his conscious mind is focused elsewhere (getting back into the "World") but his unconscious mind and autonomic nervous system are still in high alert, <u>especially</u> monitoring the external world for hints and signs of threat. They

...

simply appear edgy and may have sleep problems. They'll get over it with simple time. Right?

Wrong!

3) Unnumb: The parade is over for the rest of their lives. You get a parade for Veterans Day and 4th of July. The politicians quit coming by for "photo ops". You quit getting invited to public gathering. Other than that, you are in the silence every day, all day for the rest of your life. You don't want to go to photo ops and give speeches any more. The defense mechanisms the military gave you as part of your preparation to do these traumas to others and receive these traumas eventually weaken. Defense mechanisms? What defense mechanisms? Who, me? Then you are as raw as a burn victim whose internal nerves are exposed to the elements after their outer layer of skin is burned off.

The military uses many defense mechanisms to reduce their soldier's resistance and inhibitions to kill others. Here is a partial list. I saw one decades ago (during the Vietnam era) but can't find it yet.

"Rationalization"-the military gives the troops many reasons why killing another human being who is a father, son, mother or daughter 34, 340, or 3400 miles away is ok and necessary.

"Intellectualization"-the military training masks the interpersonal horror of killing someone by couching all the actions and weapons in scientific terms or pure jargon.

"Dehumanization"-it's easier to injure or kill another person if you don't know them as a person. We call them "the enemy", "japs", "nips", "gooks, "Charlie" "towelheads", and all the other nicknames coined the describe the opponent. They've got their nicknames for us, too. Therefore, we also develop weapons that can kill them 34, 340 or 3400 miles away with just the push of a button over here.

The troop's individual nicknames for each other dehumanize them and often make them sound like cool dudes. Ask your combat veteran for examples of these.

The night-vision scopes reduce the night-killings to video-game-like arcade shots. The kill is just more points toward getting the fantasy R'nR, isn't it?

"Sanitization"-each side in each war develops many "catchy" works to describe their weapons, their operations and the plain-old-killing involved in war. Any name or nickname of any weapon is a good example.

A weapon that is an upgraded version of our civil-war's Gatlin gunmounted in a slow-moving transport plane gets labeled "Puff the Magic Dragon" in the Vietnam era to romanticize it (with a tinge of sarcasm due to use of an anti-war singing group (Peter, Paul and Mary) song title.

It's now called a "Specter" gunship to make it sound exotic. It can kill lots of people at one time. It can put one .50-calibre round in each square yard of a football field each second it fires. Pretty cool name for a pretty cool weapon, huh? Not if you are on the receiving end.

An upgraded jeep that can climb up very steep inclines, carry several troops to their death at 65 mph, was so thin-skinned that an AK-47 round would go all the way through it during the Battle of Mogadishu until they upgraded their armor in OIF (much less the RPGs and IED's blowing them to pieces) got called the HumVee. Its civilian version got labeled the "Hummer". Cute names, huh?

The 15,000 pound bomb that has a minimal kill radius of 300-900 feet (FAS, Military Network) gets labeled "The Daisy Cutter".

The airplanes all have cool, dangerous sounding nicknames. The Mustang, Tomcat, Hellcat, Phantom, Sabre, Supersabre…on and on. All of the combat planes have them.

…

Putting a lethal round into as many of the enemy as possible gets called "Taking Care of Business", "Earning Your Paycheck", "Kicking Ass", "Taking Care of the Bad Guys", to name a few. It's all just a near-random, lethal dance of anger until someone quits.

Minimization-those previous phrases ("taking care of business", etc.) are also minimizations (defined by me as "describing the complex act in simple, understated terms"). The WWII boys were notorious for it (storming the machine guns at Omaha beach was "bad". These are all generic understatements. However, I have also heard specific understatements that mask the veterans physical and mental pain. For example, one WWII trooper who was on some of the first waves at Omaha Beach on D-day, was wounded but survived and told his family he just got "grazed" by a bullet but went on to become a spy.

After he got back home, he would go around doing the Nazi salute and shout, "Heil, Hitler". He would spit in other people's beer at the bars and then finish it off because they wouldn't want to touch it after he contaminated it. He would sell all of the furniture in the house for more liquor. He could throw things and hit his children from many yards away with anything he could get his hands on if they got noisy.

Many years after his death, I worked with a son and his family and started inquiring about his father's military experiences. A family friend who knew the father before and after the war said he came back very changed. She also said he had one eye blown out by an explosion on Omaha beach, stuck it back in and continued to fight until his disability (brain damage) was noticed.

A simple little bullet grazing his head didn't cause his bizarre actions. They were due to extensive but undocumented brain damage severe enough to blow one eye out of his head.

So, anyway, then the troops pack up, load up and ship out. They go back to the states, back to parents, back (they hope) to wives and girlfriends, back to kids, back (they hope) to jobs, back (they think) to peace and quiet. They get the band.

When the music stops and when those defense mechanisms aren't needed any more, the memories, thoughts, feelings about the soldier's experiences come creeping, flashing and flooding back. They become

UNNUMB!!!!!!!!!!

The defense mechanisms fade and crumble, tensions build, the "triggers" go off, the emotional damage gets felt for the first time. I couldn't think of a better until the other day. That word is

RAW!!!!!!!!!!!!!!

Once your defenses drop down because you think you are safe, it's like a layer of skin has dropped off your body. You are as emotionally raw as a person is who has had their skin burned off. For both of you, the slightest breeze causes excruciating pain.

I find it so tragic and ironic that the past catches with trauma victims most often when it is quiet, when they **really are** safe from further harm <u>from that arena</u>.

Then the "working through" begins. Much of it involves PTSD; much of it involves grief, guilt and shame.

"Denial"-There's the notion that sometimes a bullet "has your name on it" and there's nothing you can do. That's denial. If you look and listen to the Vietnam era movies, you hear troops saying, "It's nothing", "It don't mean a thing" and other phrases that the real combat vets have to say to themselves to try to keep from breaking down in combat or on patrol.

I heard another, much more blatant one on a video produced by a combat trooper in Iraq aired on Military Channel sometime in the week or two before August 1, 2007. He and his unit were providing protection for a privately owned supply company in Iraq. They were trying to protect four-mile-long convoys from

IEDs, snipers, RPGs and "insurgents" attacks. He narrated that they get hit by at least one IED per trip. His denial phrase was something like, "You just have to tell yourself they aren't going to get you today". That's denial.

If anyone knows the name of that film and the trooper, please email me so I can give him credit.

Patience Mason has written about five stages of post-return adjustment. I have no serious disagreement with her five versus my four stages. I've inserted hers below:

Stage 1: I'm fine: Most soldiers come back believing it's all over. Young, strong, proud, even if you are having some odd moments, you are not about to tell the doctors because you will be kept from going home. The changes that helped you survive war don't seem that big a deal, and who is going to tell some guy in a white coat that you are seeing dead people? You may not know how much you have changed till you're home. Life here is flat. People have petty problems. You can't sleep, have bad dreams, get furious at everything, and keep looking for roadside bombs. When a car backfires, you hit the dirt. Still, you probably think the people around you have problems. Not you. Any comments about how you've changed may really piss you off. You're fine! You survived a war! What kind of help could you possibly need after that? If you don't know that it is normal to be affected, what else can you do but deny that you are? That's what everyone else does. Denial can make your family feel nuts. You may be telling them they are nuts. This usually does not improve relationships. Furthermore, in today's military, you probably will have to go back, so denial may seem necessary.

Note: She spelled "fine" as "f-i-n-e". In military jargon that stands for "fucked up, insecure, neurotic and emotional".

Stage 2: I'm not fine, but I'm not telling you: You notice some problems. You get angry too fast, you are yelling at people instead of talking to them, you keep seeing your friends die. When civilian things go wrong, you don't care. (Is anyone shooting?) You may be shocked to feel nothing when a beloved relative gets sick or dies, or you may think you don't love your spouse anymore because you can't feel it. You hate civilians or Arabs. You are not fine, but you are not going to tell anyone, especially not anyone who wasn't there and has been telling you that you have problems. You start to think that you can't talk to anyone who wasn't there. You begin isolating so no one will see how nuts you feel. You are pissed off about being affected. You also fear going for help because it may dull your edge, which you will need when you go back. It might also affect your career, and you don't want people to think you are nuts. You exclude your spouse. He or she gets angry at you a lot.

Stage 3: I can't talk to people who weren't there: Since you can talk to other vets, you feel that no one understands unless they were there. This unfortunately is true. Most people make this clear by saying insensitive things like, "So what's your problem? Get over it!" "Did you kill anyone?" "You're a hero." and the inevitable, "But why aren't you over it?" So you increase your isolation from family and friends. This however tends to make spouses angry, because we are supposed to be understanding. Your sense of humor has become very black, and you laugh at things that would have horrified you once. You may even wonder if your spouse would still love you if they knew what happened over there. You might feel that everyone around you is spoiled and insensitive and it pisses you off. You have to stay so numb that your spouse feels you don't love him or her anymore.

Stage 4: What's wrong with me? The term "Post-Traumatic Stress Disorder" is a good description of the effects of war on normal people except it is not a disorder when you are under fire. The skills of war create a lot of disorder in your civilian life. Shrinks and family members tend to see the symptoms of PTSD as the problem. Not me. I see war as the problem and the symptoms of PTSD as solutions to the problem of war, something right with you, not something wrong with you. Each symptom begins as part of your body's hard-wired survival responses to danger, which your training has been designed to intensify and strengthen. They worked. You are alive. That is the bottom line. You have been through something that killed others. Having PTSD is proof of survival. I also believe that the people

...

who get PTSD are the ones who care the most. You may feel like you don't care, but if you didn't care, you would not have to develop the symptom of emotional numbing to survive. Although PTSD symptoms originate in hard-wired survival skills built into all of us, unhealed, they can become your biggest problems over time.

Stage 5: I'm screwed up and no one can help. Deciding that no one can help is pretty human, but it is not true. I don't think you are screwed up, either. You are in survival mode. What helped you survive one deployment will probably help you during the next one, unless your symptoms become debilitating. What you will need, when you are finally home for good, and decide you want to change, is information and tools, someone to talk to, and hope. So will your family. You can get treatment without diagnosis at the Vet Centers and for five years after you get back at the VA. This can be a problem since it might be two years before you realize you have problems.

I think she's totally right on.

Hypothesis 2: The wounded now are worse off than ever before.

I've lectured about PTSD for several decades. Early in my career, I coined a phrase to describe the agony of the traumatized that didn't work right ("The lucky ones only get killed") and it was insensitive and too clumsy to convey that real message. A Vietnam veteran friend of mine recently readjusted it to the following:

"The lucky ones in war don't get wounded"

That, too, doesn't have much pizzazz to it because it's too obvious. Getting wounded is a bad thing. All soldiers who don't get wounded feel fortunate if not downright lucky. I'll work on it and get back to you if I get some better phrase.

The real issue I want to convey is "The wounded (since Vietnam) are probably worse off than the wounded in previous wars because they have such a higher (physical) survival rate from severe wounds (that will lead to greater physical and emotional suffering) compared to soldiers the fought and got wounded in previous conflicts. The wounded ones in pre-Vietnam days were more likely to simply die. The Vietnam and after wounded will need more attention for their physical and emotional wounds than the pre-Vietnam troops thanks to a) the helicopters, and b) our vastly improved understanding and treatment of physical traumas.

The WWII combat vets hoped for the "Million Dollar Wound". It was a wound that wasn't bad enough to kill or severely maim you but was bad enough that they wouldn't send you back into combat and sent you home.

Helicopters were nonexistent in WWII operations. They were used in military action during the Korean War, but primarily for medical transport. There weren't many of them compared to their presence in Vietnam and hereafter. They get wounded troops out of the hostilities much quicker after the wounds and (often) with much less trauma than the WWII open jeeps bouncing along muddy, rutted roads.

What is the wounded/wounded-that-died ratio? All numbers are suspect but I think the Department of Defense's own numbers tell an adequately accurate story of that ratio for sake of my issue. The ratio for both the Civil War and WWII was between 4:5 and 1: 2 That means that almost four out of five in the Civil War and half of those in WWII that got wounded eventually died. Vietnam's ratio was 1: 5. That means that only one out of five wounded died. The current conflict's ratio is even "better" (1:8). This means that only one out of eight wounded eventually died. The chart below gives some raw numbers.

...

Combat-Wounded

Civil War: 43,012 (source: Fox's Regimental losses, Chapter 2 at http://www.civilwarhome.com/foxschapter2.htm

WWI: 20 million wounded Source: http://encarta.msn.com/encyclopedia_761569981/World_War_I.html

WWII: 671,801: source: **World War II Factbook** @http://www.skalman.nu/worldwar2/stupade.htm

Korean War: 79,526

Vietnam: 304,000 (75,000 "severely disabled")
Source: STATISTICAL DATA ON STRENGTH AND CASUALTIES FOR KOREAN WAR AND VIETNAM @ **http://www.army.mil/cmh-pg/documents/237adm.htm**. Statistics about the Vietnam War @ http://www.vhfcn.org/stat.htm

Iraq 27,000 (as of 7/24/07) Source: CNN

However, there is a flip-side to the higher survival rate. The severely wounded who would have died in pre-WWII conflicts survive now but have serious and life-long physical and mental disabilities. As with all physical deterioration, the medical profession can sustain duration of life but quality of life is so poor as to be "poor" (I call it grotesque, degrading, depressing, demoralizing and sadistic) at the end for many.

By the way, many of the surviving elder also don't opt to continue living in that state of misery because the highest rate of suicide (in America) is in the elder, especially college-educated, Caucasian males.

There are many articles available on line if you want to look at this issue further. For example, refer to Ann Scott Tyson's Washington Post article (U.S. Casualties in Iraq Rise Sharply: Growing American Role in Staving Off Civil War Leads to Most

Wounded Since 2004http://www.washingtonpost.com/wp-dyn/content/article/2006/10/07/AR2006100700907.html for further discussion of this issue. Or simply search "wounded/killed" ratio.

The conclusion that many of these writers (and I) come to is that the better ratio now doesn't mean there is less misery for the survivor. I just don't want the average viewer of the CNN shows such as "Combat Hospital" (or the surviving veteran and their family) to think their battle ends when they come back home. I don't want those surrounding the surviving veteran to think he has it better with his wounds (the visible or invisible).

It, unfortunately, can and will be a life-long (for all wounded), life-losing struggle (for some) if they don't get help from their invisible wounds.

If you don't know the agony that long-term survivors of most serious physical injuries endure, go to the Orthopedic rehab unit of your nearest hospital. Visit with the head and spine injured. Follow their progress (or decay) for 10 years.

Their pain gets worse, their tolerance for pain gets worse, their need for pain medication increases, their addictions to pain medications always get worse (because all pain medications are addictive), their use of illegal drugs often gets worse, they often get clinically depressed (because losing control of your body and having chronic pain is truly and unavoidably depressing), and their blatant suicide or accidental suicide is high.

IS THERE A GENETIC COMPONENT TO SUSCEPTIBILITY or "RESILIIENCE" TO PTSD?

I did not initially have time to write enough detail about this new view of PTSD. I researched it for the second edition, put it down in writing but cannot find that version. Therefore, I revert to my initial writings and assure the reader there is scientific basis for my decision:

...

I am actually offended by those that suggest that some races are "genetically vulnerable" to being less able to respond to or cope with trauma which also infers that other races are genetically superior to coping with trauma.

There have been many scientific studies published that suggest there is some genetic resilience or toughness against being traumatized. Conversely, they have suggested some races are more genetically vulnerable to being traumatized. The earliest studies on resilience to trauma were done by Israeli scientists (Hobfoll, et al., 2006; Hobfoll et al., 2009; Johnson, et al., 2009). They compared the amount of PTSD in Israelis and Palestinians after terrorist attacks in 2004-2005. They found that Israelis in the conflict reported fewer PTSD symptoms that the Palestinians and concluded that Israelis were genetically superior to Palestinians in coping with trauma. This study's results are highly suspect in that the researcher's race (Israeli) could have influenced the outcome of their data (self-fulfilling prophecy, also called the Rosenthal Effect or Experimenter Bias). This study's results are also highly suspect in that the experimenters knew the race of each subject, also potentially biasing their overt or covert interview/data collection technique. Thirdly, the study's results are suspect because the Israelis won that conflict and the outcome of a trauma might influence the perception of the severity of the trauma and/or the meaningfulness of the trauma.

Another study by Rueff et al. (undated) that suggested a genetic relationship to the ability to tolerate stress and trauma was done comparing African-American, Hispanic and Caucasian Vietnam War veterans on the level of PTSD they experienced. This study found that Hispanic Vietnam war veterans suffered more PTSD symptoms and concluded that Hispanics were genetically inferior to tolerating stress or trauma. The study did not take into account any other factors. I suggest that those authors should have more precisely assessed for the amount of tissue damage the veterans in this study witnessed since mere presence of PTSD symptoms does not fully tell you how severely they have been traumatized. The study also did not account for the higher emotional investment Hispanic soldiers had compared to some of the other racial groups in the study since Hispanics viewed military service as a way into greater economic standing and societal prestige in American society than was available to them through the most common job for them at that time, manual labor.

In more recent years, King (2013, p. 45) wrote that there is a "genetic predisposition" to PTSD and cited Mehta & Binder, 2012) and Skelton et al., 2012) as supporting studies, Mehta & Binder actually wrote "studies investigating main genetic effects associated with PTSD have yielded inconsistent findings" and "a combined analysis of environmental, genetic, endophenotype and epigenetic data will be necessary to better understand pathomechanisms in PTSD." (both quotes taken from the Abstract). King's other quoted source, Skelton et al, also stated "The majority of the initial investigations into main effects of candidate genes hypothesized to be associated with PTSD risk have been negative..." (also in their abstract). Therefore, King's bland assertion that resilience to PTSD "is genetic" is grossly simplified and misleading.

Two psychiatric geneticists recently cautioned the rush to belief of a genetic cause when they wrote, "Psychiatry as a discipline has too often been characterized by many speculations based on few facts". (Zerbin-Rudin and Kendler, 1996). Joseph (2004) more specifically cautions against accepting the findings of genetic causes for any mental disorder due to two different factors. First, most researchers begin by stating that prior studies results have established the role of genetics in the mental disorder of issue. This blind assumption can easily lead the researcher into committing The Rosenthal Effect (when the researcher's assumptions bias the results and conclusions). For example, Tsuang and Faraone (200) wrote, "A century of genetic epidemiologic research shows that genes play a substantial role in the etiology of schizophrenia. This is the only reasonable conclusion we can draw from family, twin and adoption studies". Both Dr. Joseph's book and my own (*The Inherited Insanity Illusion*) clearly and completely show that there are many environmental factors related to *any* mental disorder. I refer the interested reader to both of these books because the topic is too complex to discuss adequately and comprehensively in this article.

The second problem mentioned by Joseph (2004) is that both methods of molecular genetic research (linkage and association) have serious flaws.
With the first method (linkage), researchers look for genetic markers linked to the alleged disease among consanguineous (blood-related) family members. These linkage studies are designed to identify areas of chromosomes where relevant genes might be located but are unable to identify actual genes. That is the task of follow-up studies. Finding a linkage between gene area and behavior does not mean the gene has been found. For example, Colbert (2001) said finding a marker and concluding it is the cause of a behavior is similar to a prospector finding a trace of gold and assuming there is a vein of gold nearby.

Association studies compare the frequency of genetic markers among unrelated affected individuals and a control group.
In both types of these studies, any positive findings are later retracted publically or fail to be replicated (which are usually not publically reported at all). The classic and most glaring example of failure to replicate was Sherrington et al.'s 1988 article published in the journal *Nature* which had a failure to replicate the same results by Kennedy et al. in the same article. Egeland et al. (1987) also published but had to withdraw a study initially reporting a genetic basis for bipolar disorder.
Another "watershed" study (DeLisi et al., 2002) studied 382 sibling pairs but failed to replicate reported linkages in several previous studies. These investigators wrote that the most striking feature of their results "is a failure to confirm a number of earlier claims of positive findings" and "the present findings suggest that a critical reevaluation of the linkage approach is warranted."

As prestigious an institution as the American National Institute of Health concluded this about the role of genetics in mental illness: "Although statistically significant, each of these genetic associations individually can account for only a small amount of risk for mental illness,". Because of this, the variations couldn't yet be used to predict or diagnose specific conditions." (2013).

While laboratory research on PTSD is one area for data, real life, real death, real life-or-death situations strike me as above and beyond the real possible in the laboratory. Let me put it most bluntly and grotesquely: Any person of any race who gets a sharpened, bamboo stick shoved up their anus will be equally traumatized, no matter what race he/she is. All other physically damaging assaults will produce equal massive physiological and psychological traumas no matter what race the recipient is. While there are other factors that influence the expression of the trauma after it is over, these are all societal and socially related, such as unemployment, alcohol use, social support, and stressful (nontraumatic) life events (Possemato et al., 2014). These relevant traumas can even be the many prenatal and postnatal physical stressors that produce long-term mental problems mentioned by Nolen, 2013.

...

BIBLIOGRAPHY

Agent Orange Home Page: www1.va.org/agentorange

Army withholding findings of Madigan PTSD probe @http://news.ca.msn.com/top-stories/army-withholding-findings-of-madigan-ptsd-probe (accessed 3/1/13).

Combat Clips: A Selection of OEF/OIF Veteran Statistics, October 2009 (http://ptsdcombat.blogspot.com/2009/10/combat-clips-selection-of-oefoif.html)

Fox's Regimental losses, Chapter 2 at http://www.civilwarhome.com/foxschapter2.htm

Newer Antipsychotic Drugs Raise Fatal Heart Attack Risk by: David Gutierrez@http://www.naturalnews.com/025724.html, 2009

Springfield (Missouri) News-Leader, August, 2000.

STATISTICAL DATA, ON STRENGTH AND CASUALTIES FOR KOREAN WAR AND VIETNAM ON @ http://www.army.mil/cmh-pg/documents/237adm.htm

World War II Factbook @http://www.skalman.nu/worldwar2/stupade.htm, nd

Statistics about the Vietnam War @ http://www.vhfcn.org/stat.htm, ND

CNN: July 25, 2007: 27,000 wounded in Iraq

FAS, Military Network, **http://www.fas.org/man/dod-101/sys/dumb/blu-82.htm**, ND

"Killing Our Own" @http://www.ratical.org/radiation/KillingOurOwn/KOO3.html, 2009.

Scott Tyson, Ann. U.S. Casualties in Iraq Rise Sharply: Growing American Role in Staving Off Civil War Leads to Most Wounded Since 2004, http://www.washingtonpost.com/wp-dyn/content/article/2006/10/07/AR2006100700907.html (ND)

"Shell shock" videos: http://www.liveleak.com/view?i=ed6_1193152676 and @ http://www.liveleak.com/view?i=c1c_1235258924

"Suicide Wall: suicide statistics" at **http://www.suicidewall.com/SWStats.html**

World War I, Encyclopedia Article http://encarta.msn.com/encyclopedia_761569981/World_War_I.html

Iraq War Veterans Organization: http://groups.msn.com/IraqWarVeterans

Brown, Dale. Strike Force: A Novel, p. 183, 2007. Publisher unknown.

Bertel, Rosalie. Gulf War Syndrome, Depleted Uranium and the Dangers of Low-Level Radiation @http://www.ccnr.org/bertell_book.html (2009)

Casey T. Taft, Amy E. Street, Amy D. Marshall, & Deborah J. Dowdall. Posttraumatic stress disorder, anger, and partner abuse among Vietnam combat veterans. Journal of Family Psychology, Vol 21(2), Jun 2007. pp. 270-277.

...

Colbert, T.C. (2001). Blaming our genes. Tustin, Ca.: Keveco.

Common Genetic Factors Found in 5 Mental Disorders@ http://www.nih.gov/researchmatters/march2013/03182013mental.htm. Accessed 6/2/15.

Diagnostic and Statistical Manual, Third Edition, American Psychiatric Association, 1980.

Diagnostic and Statistical Manual, Fourth Edition. American Psychiatric Association, 1994.

Damasio, A.B. (1994). Descartes' error: Emotion, reason, and the human brain. New York: Avon Books.

DeLisi, I.E., Shaw, S.H., Crow. T.J., Shields, G., Smith, A.B., Larch, V.W., Wellman, N., Luftus, J. Nanthakumar, B., Razi, J. Stewart, J., Comazzi, M., Vita, A., Heffner, T., & Sherington, R. (2002). A genome-wide scan for linkage to chromosomal regions in 382 sibling pairs with schizophrenia or schizoaffective disorder. American Journal of Psychiatry, 139, 803-812.

Drescher, Kent D, Rosen, Craig S, Burling, Thomas A, Foy, David W. Causes of death among male veterans who received residential treatment for PTSD. Journal of Traumatic Stress, Vol. 16(6), Dec 2003. pp. 535-543

Egeland, J.A., Gerhard, D.S., Pauls, D.L., Sussex, J.N>, & Kidd, K.K. (1987). Bipolar affective disorders linked to DNA markers on Chromosome II. *Nature*, 325, 783-787.

Friedman, Matthew J., M.D., Ph.D. Post Traumatic Stress Disorder: The Latest Assessment and Treatment Strategies, Dean Psych Press Corp, Kansas City, Mo, 2001.

Gerardi, Robert J, Keane, Terence M. ,Cahoon, Beth J. & Klauminzer, Guy W.An in vivo assessment of physiological arousal in posttraumatic stress disorder. Journal of Abnormal Psychology, Vol 103(4), Nov 1994. pp. 825-827.

Henderson, Charles. Marine Sniper: 93 Confirmed Kills, 2001, p. 35. Publisher unknown.

Hobfoll, S.E. & Canetti-Nisson, D, & Johnson, R.J. (2006). Exposure to Terrorism, Stress-Related Mental Health Symptoms and Defensive Coping Among Jews and Arabs in Israel. Journal of Consulting and Clinical Psychology, 74(2), 207-218.

Hobfoll, S.E., Palmierii, P.A., Johnson, R.J., Canetti-Nisim, D., Hall, B.J. & Galea, S. (2009). Trajectories of Resilience, Resistance, and Distress During Ongoing Terrorism: The Case of Jews and Arabs in Israel. Journal of Consulting and Clinical Psychology, 77(1), 138-148.

Johnson, R.J., Canetti, D., Palmieri, P.A., Galea, S., Varley, J. & Hofboll, S.E. (2009). A Prospective Study of Risk and Resilience Factors Associated With Posttraumatic Stress Symptoms and Depression Symptoms Among Jews and Arabs Exposed to Repeated acts of Terrorism in Israel. Psychological Trauma: Theory, Research, Practice and Policy 1(4), 291-311.

Jones, E. & Thomas, A (2007). Shell shock: An outcome study of a First World War 'PIE' unit. Psychological Medicine, Vol 37(2), Feb 2007. pp. 215-223.

Kennedy, J. (2015). Vets Can Finally Sue Contractors for Cancer Caused by War. **http://www.thedailybeast.com/articles/2015/02/02/possible-relief-for-vets-sick-from-war-zone-burns-pits.html?utm_content=buffer8f0f0&utm_medium=social&utm_source=facebook.com&utm_campaign=buffer**. Accessed 4/26/15.

King, L.A. (2013). Experience Psychology, second edition. McGraw-Hill, NYNY.

...

Lunde, D.T. Murder and Madness, Scribner's Sons (NY) and W.H..Freeman (San Francisco, 1975)

Mehta D. & Binder, E.B. (2012) Gene environment vulnerability factors for PTSD: The HPA-axis. *Neuropharmacology*, 62,654-662.

Nolen, F. (2014). *Teratogens and TBIs: the Silent Slaughters.* Amazon.com.

Pitman, R. K., Orr, S. P., Forgue, D. F., de Jong, J. B., & Claiborn, J. M. (1987). Psychophysiologic assessment of posttraumatic stress disorder imagery in Vietnam combat veterans. Archives of General Psychiatry, 44, 970–975.

Possemato, K., McKenzie, S., McDevitt-Murphy, M. Williams, J. L. & Ouimette, P. (2014). The relationship between postdeployment factors and PTSD severity in recent combat veterans. *Military Psychology*, Vol 26(1), Jan, 2014. pp. 15-22

Ruef, A.M., Litz, B.T., & Schlenger, W.E.(?). Hispanic Ethnicity and Risk for Combat-Related Posttraumatic Stress Disorder. Cultural Diversity and Ethic Minority Psychology, 6(3), 235-251).

Sherrington, R., Brynjolfsson, J. Petursson, H., Potter,M., Duddelston, K. Barraclough, R., Wasmuth, J., Dobbs, M., and Gurling, H. (1989). Localization of a susceptibility locus for schizophrenia on chromosome 5. *Nature.* 336, 164-167.

Skelton, K. Ressler, K.J., Norrholem, S.D.,Jovanovic, T. & Bradley-Davino. B. (2012) PTSD and gene variants: New pathways and new thinking. *Neuropharmacology*, 62, 628-637.

Survivor Guilt at
www.spike.com/collection/15805?cmpnid=800&lkdes=COL_15805, 2009.

Tsuang, M.T. & Faraone, S.V. (2000). The genetics of mood disorders. Blatimore: Johns Hopkins University Press.

Zoroya, G. OEF/OIF Veteran Suicide Toll: Nearly 15% of Overall U.S,. Military Casualties Result from Suicide. PTSD Combat: Winning the War Within, September 15, 2008, at http://ptsdcombat.blogspot.com/2008/09/oefoif-veteran-suicide-toll-15-of.html, accessed 10/08.

Zerbin-Rudin, E. & Kendler, K.S. (1996). An introduction to their family studies of schizophrenia. American Journal of Medical Genetics, 67, 332-337.

...

About the Author

Dr. Nolen has a BA, MA and Ph.D. from the University of Missouri-Columbia. He was recognized as an expert witness by the courts of Missouri on combat PTSD in 1986 and recognized an expert on Child Abuse: Victims and Perpetrators in 2005.

He has lectured extensively nationwide on child abuse and neglect: victims and perpetrators, Dissociative Identity Disorder (formerly known as Multiple Personality Disorder), Traumatic Brain Injury and Combat-Induced PTSD.

He has produced CEU home-study courses on all of those topics above plus three others that were sponsored by nation-wide CEU companies (see list below).

He has published books on all six of those issues.

He's had a private practice since 1975 and has taught at colleges and universities in Missouri, Kentucky and Indiana.

Other Books by the Author

A Dissociative Identity Disorder Casebook: Diagnosis, Treatment and Case Management. Amazon.com. 2009.

The Inherited Insanity Illusion. Amazon.com. 2011

Teratogens and TBIs: the Silent Slaughters. Amazon.com. 2014

Educator Sexual Perpetrators. Amazon.com. 2016

Factitious Disorders: Facts and Fantasies. Amazon.com, 2015.

Made in the USA
Las Vegas, NV
13 April 2026

45272437R00098